Recent Advances in High-Power Electromagnetics

Other related titles:

You may also like

- Wyatt | EMI Troubleshooting Cookbook for Product Designers (Electromagnetic Waves) | 2014
- James B. Tsui | Microwave Receivers with Electronic Warfare Applications | 2005
- Kai Borgeest | EMC and Functional Safety of Automative Electronics | 2018
- Guillaume Andrieu | Electromagnetic Reverberation Chambers: Recent advances and innovative applications | 2020

We also publish a wide range of books on the following topics:
Computing and Networks
Control, Robotics and Sensors
Electrical Regulations
Electromagnetics and Radar
Energy Engineering
Healthcare Technologies
History and Management of Technology
IET Codes and Guidance
Materials, Circuits and Devices
Model Forms
Nanomaterials and Nanotechnologies
Optics, Photonics and Lasers
Production, Design and Manufacturing
Security
Telecommunications
Transportation

All books are available in print via https://shop.theiet.org or as eBooks via our Digital Library https://digital-library.theiet.org.

ELECTROMAGNETIC WAVES SERIES 570

Recent Advances in High-Power Electromagnetics

Edited by
Felix Vega, Nicolas Mora, Edl Schamiloglu and
Chaouki Kasmi

The Institution of Engineering and Technology

About the IET

This book is published by the Institution of Engineering and Technology (The IET).

We inspire, inform and influence the global engineering community to engineer a better world. As a diverse home across engineering and technology, we share knowledge that helps make better sense of the world, to accelerate innovation and solve the global challenges that matter.

The IET is a not-for-profit organisation. The surplus we make from our books is used to support activities and products for the engineering community and promote the positive role of science, engineering and technology in the world. This includes education resources and outreach, scholarships and awards, events and courses, publications, professional development and mentoring, and advocacy to governments.

To discover more about the IET please visit https://www.theiet.org/.

About IET books

The IET publishes books across many engineering and technology disciplines. Our authors and editors offer fresh perspectives from universities and industry. Within our subject areas, we have several book series steered by editorial boards made up of leading subject experts.

We peer review each book at the proposal stage to ensure the quality and relevance of our publications.

Get involved

If you are interested in becoming an author, editor, series advisor, or peer reviewer please visit https://www.theiet.org/publishing/publishing-with-iet-books/ or contact author_support@theiet.org.

Discovering our electronic content

All of our books are available online via the IET's Digital Library. Our Digital Library is the home of technical documents, eBooks, conference publications, real-life case studies and journal articles. To find out more, please visit https://digital-library.theiet.org.

In collaboration with the United Nations and the International Publishers Association, the IET is a Signatory member of the SDG Publishers Compact. The Compact aims to accelerate progress to achieve the Sustainable Development Goals (SDGs) by 2030. Signatories aspire to develop sustainable practices and act as champions of the SDGs during the Decade of Action (2020–2030), publishing books and journals that will help inform, develop, and inspire action in that direction.

In line with our sustainable goals, our UK printing partner has FSC accreditation, which is reducing our environmental impact to the planet. We use a print-on-demand model to further reduce our carbon footprint.

websites and does not guarantee that any content on such websites is, or will remain, accurate or appropriate.

Whilst every reasonable effort has been undertaken by the Publisher and its licensors to acknowledge copyright on material reproduced, if there has been an oversight, please contact the Publisher and we will endeavour to correct this upon a reprint.

Trade mark notice: Product or corporate names referred to within this publication may be trade marks or registered trade marks and are used only for identification and explanation without intent to infringe.

Where an author and/or contributor is identified in this publication by name, such author and/or contributor asserts their moral right under the CPDA to be identified as the author and/or contributor of this work.

British Library Cataloguing in Publication Data
A catalogue record for this product is available from the British Library

ISBN 978-1-83953-947-3 (hardback)
ISBN 978-1-83953-948-0 (PDF)

Typeset in India by MPS Limited

Cover image credit: Foto-Ruhrgebiet via Getty Images

Contents

About the editors

Felix Vega is the chief researcher at the Directed Energy Research Center (DERC); a Center affiliated with the Technology Innovation Institute (TII), in Abu Dhabi, UAE. He oversees the applied research and development programs on high-power electromagnetics, EMC, high Energy lasers, and sensing systems. He is secretary of the IEEE Antennas and Propagation Society and associate editor of the *IEEE Transactions on Antennas and Propagation*.

Nicolas Mora is an assistant professor in the Department of Electrical and Electronics Engineering at Universidad Nacional de Colombia, Bogota, Colombia. In 2018, he received the HPEM Fellow award from the Summa Foundation, and in 2019, the Motohisa Kanda Most Cited IEEE Transactions in EMC Paper Award. Since 2022, he has been associate editor of the *IEEE Transactions on Electromagnetic Compatibility*. In 2023, he joined the Board of Directors of the IEEE EMC Society as a representative of R9.

Edl Schamiloglu is a distinguished professor of electrical and computer engineering at the University of New Mexico, USA. He is also director of the Directed Energy Center at the University of New Mexico (DEC@UNM). He has co-authored over 500 refereed journal and conference papers, and nine patents. His publications have been cited over 10,000 times. Professor Schamiloglu is a fellow of the IEEE, a fellow of the American Physical Society, and an EMP fellow (sponsored by the Summa Foundation).

Chaouki Kasmi is the chief innovation officer at the Technology Innovation Institute (TII), Abu Dhabi, UAE. Dr. Kasmi has co-authored more than 200 scientific papers. He also is an associate electromagnetics scientist and researcher at the Faculty of Electrical Engineering at the Helmut Schmidt University. He serves as a vice president of the Commission E French Section and vice chair of the International Commission E of the International Union of Radio Science (URSI). His scientific interests are the R&D of Directed Energy systems, Radars and Electro-optics.

Chapter 1

Recent advances in high-power electromagnetics

*Felix Vega[1], Nicolas Mora[2], Edl Schamiloglu[3]
and Chaouki Kasmi[1]*

This work presents a selection of contributions presented at the GlobalEM 2022 Conference supplemented by a selection of unpublished studies from key experts from the high-power electromagnetics (HPEM) community. It provides a discussion on the state-of-the-art in high-power electromagnetics (HPE) technologies, focusing on high-power microwave and ultrawideband sources, intentional electromagnetic interference (IEMI), electromagnetic compatibility (EMC), and detection methods for IEMI threats.

In what follows, we offer a short overview of the current scientific and industrial research and development activities.

1.1 High-power electromagnetics today

High-power electromagnetics (HPEM) is a field that is continually evolving and has far-reaching applications across both military and civilian domains. Traditionally, the field has been characterized by infrastructure-based HPEM sources that require substantial fixed installations with extensive support systems [1]. While powerful and effective, these sources are often limited by their lack of mobility and the significant infrastructure they demand.

The field is undergoing a transformative shift towards developing deployable and transportable sources that, although offering a fraction of the performance of laboratory systems, can be conveniently integrated into mobile platforms or transported to the operational environment.

Narrowband systems primarily rely on vacuum electronic technologies, both classic and relativistic. Examples include relativistic magnetrons, pulsed magnetrons, backward wave oscillators (BWOs), and others [2]. These technologies have been foundational in developing high-power microwave (HPEM) sources capable

[1]Directed Energy Research Center, Technology Innovation Institute, Abu Dhabi, United Arab Emirates
[2]Department of Electrical and Electronics Engineering, Faculty of Engineering, Universidad Nacional de Colombia, Bogota, Colombia
[3]Department of Electrical and Computer Engineering, University of New Mexico, USA

of delivering hundreds of Megawatts to tens of Gigawatts of power during tens of nanoseconds [2].

Ultrawideband systems, on the other hand, have been predominantly based on pressurized spark gap switch technologies. These systems can generate short pulses with sub-nanosecond rise times and amplitudes of several [3,4] covering a broad spectrum of frequencies. This makes them particularly useful for disabling electronic systems over a wide range of frequencies [4,5].

In the open literature, successful versions of both narrowband and ultrawideband sources have been reported, at least at the demonstration level, showcasing their potential for practical implementation [6,7].

On the other hand, while not reaching the power levels of traditional vacuum electronic sources, solid-state devices offer advantages that make them appealing for specific applications. One key benefit of solid-state technology is its portability. Unlike vacuum technology, solid-state devices are compact and lightweight, making them ideal for mobile and field-deployable systems. This portability is particularly beneficial in applications requiring mobility, such as systems requiring rapid field deployment. Additionally, solid-state devices are more efficient in terms of power consumption, requiring less energy to operate. This efficiency translates into longer operational times and reduced energy. Furthermore, their reduced need for ancillary systems, such as large cooling units and complex power supplies, simplifies their integration into various platforms and lowers the overall system costs.

1.2   HPE in counter UAV applications

The increasing presence of UAVs in operational scenarios has introduced new security challenges worldwide, especially with the proliferation of COTS (Commercial Off-The-Shelf) and modified drones.

Initially, COTS amateur drones were primarily used for surveillance and intelligence gathering, capturing video and photographs. However, some models were soon modified to deliver small payloads such as explosives or grenades and chemical or biological agents. Reports include incidents where these drones have been used for such purposes.

A new trend has emerged involving using DIY (Do-It-Yourself) custom models. These are assembled using critical components purchased online, such as motors and electronic control boards, while other parts, like the fuselage and payload, are sourced locally. These DIY drones present unique challenges due to their customization and potential for advanced functionalities.

Customized drones can be manufactured at the prototype level, where one or a few units are produced for specific purposes. They can also be produced on a medium-to-large scale, resulting in single-mission, low-cost drones deployed in swarms to overwhelm air defenses.

HPE sources can be integrated into a layered counter-UAV system, including jamming and electromagnetic disruption.

Jamming disrupts the radio link used for control and telemetry, severing communication between the operator and the drone. This interference can cause the drone to lose its navigational commands, potentially forcing it to land, return to its takeoff point, or hover until the signal is restored. The control channels typically operate on ISM (Industrial, Scientific, and Medical) frequencies, which include 2.4 GHz for remote control and 5.8 GHz for video transmission. Long-range control frequencies are often 433 MHz and 915 MHz for extended control and telemetry communications. Some advanced drone models also utilize links at 1.3 GHz for specific applications. Due to these precise frequencies, waveform, power control, and agility requirements, jamming can only be effectively achieved using transistor-based RF sources [8,9].

The second approach involves using HPEM sources to induce noise and disturbances within the drone's circuitry, effectively neutralizing it. This method disrupts the drone's internal electronics, causing it to malfunction or become inoperable. A single, land-based high-power system can achieve this, reaching an effective range of up to 2 km [10]. These systems are typically stationary and can cover large areas, making them suitable for protecting critical infrastructure. However, one drawback of stationary solutions is their lack of mobility, making repositioning them quickly in response to moving threats challenging.

Alternatively, smaller HPE systems can be mounted on interceptor drones. These mobile platforms can actively seek out and approach target drones, activating the electromagnetic source within a range of tens of meters [11]. For this approach to be effective, the interceptor drone must be electromagnetic compatibility (EMC) hardened to prevent self-disruption during the operation. Nevertheless, a drawback of this approach is the limited operational range and endurance of interceptor drones, which may restrict their ability to sustain long missions or engage multiple targets over an extended period.

To maximize effectiveness, it is crucial to study the targeted drone's vulnerability to electromagnetic pulses, ensuring the chosen countermeasure is appropriate for the specific drone model. This involves conducting detailed analyses and assessments of the drone's design, materials, and electronic components. Understanding the drone's frequency ranges, control systems, and shielding techniques can help identify potential weak points where electromagnetic disruption would be most effective.

Electromagnetic radiation can enter a drone either through front door coupling, where the interference enters through the drone's antenna systems, or back door coupling, where it penetrates through unintended pathways, such as seams, cables, or other openings in the drone's structure. Additionally, simulations and testing in controlled environments can provide valuable insights into how different models respond to various levels of electromagnetic interference. By tailoring the countermeasures to exploit specific vulnerabilities, it becomes possible to enhance the precision and reliability of the intervention, thereby increasing the likelihood of successfully neutralizing the threat posed by unauthorized or hostile drones. This targeted approach not only improves the efficiency of the counter-UAV systems but also helps conserve resources by avoiding the deployment of overly generalized or ineffective countermeasures [12].

1.3   Current industrial applications

High-power microwaves (HPEM) and directed energy have seen significant adoption across various industrial sectors due to their efficiency and unique capabilities.

In the pharmaceutical industry, HPEM is employed for synthesizing active pharmaceutical ingredients (APIs) and drying and sterilizing equipment and products. These applications enhance reaction rates and improve yield while ensuring sterility [13,14]. In the medical field, HPEM technology is utilized for therapeutic and diagnostic purposes. For example, microwave ablation is a minimally invasive cancer treatment, and microwave imaging techniques are used for diagnostics [15–17].

The oil industry uses HPEM for enhanced oil recovery processes [18], like microwave-assisted pyrolysis [19] and desulfurization [20], which boost efficiency and reduce the environmental impact of oil extraction and processing. In the automotive sector, HPEM is applied in the manufacturing and processing of composite materials, curing adhesives, and drying coatings and paints, improving product quality and shorter processing times [21,22].

One of the sectors where the future adoption of HPEM, especially millimeter (mm)-wave high-power sources, holds great promise includes high-efficiency drilling, which uses HPEM generators to enhance drilling processes in industries like oil and gas, mining, and geothermal energy. This technology allows cost-effective deeper penetration of the earth's crust. Estimates suggest that while the cost of mechanical drilling increases exponentially with depth due to rock-induced stress, the cost of drilling using mm-wave technologies increases linearly with depth [23].

Microwave welding is a technique in materials engineering that utilizes microwave energy to join materials through dielectric heating. This method is particularly effective for thermoplastic composites, ceramics, and certain metals, as it provides uniform heating and minimizes thermal gradients, which can lead to defects in traditional welding processes. Current research underscores the efficiency of microwave welding in producing strong, durable bonds with minimal thermal damage to adjacent materials. Applications span various industries, including aerospace, automotive, and biomedical, where precision and material integrity are critical. Advantages of microwave welding include reduced processing times, lower energy consumption, and the capability to weld dissimilar materials with minimal distortion [24,25].

Currently, microwave-based waste treatment is a promising technology for the effective management of various waste types through the process of microwave-induced oxidation. This method leverages the rapid and uniform heating properties of microwaves to facilitate the oxidation of organic and inorganic waste materials. One of the primary advantages of microwave-based waste treatment is its ability to achieve high temperatures quickly, leading to efficient decomposition and reduction of waste volume. Additionally, the process can be precisely controlled, resulting in minimal environmental impact compared to conventional thermal treatments. Microwave-induced oxidation also produces fewer hazardous bypro-ducts and can be applied to various industrial, medical, and municipal wastes. The potential for integrating microwave technology with existing waste treatment infrastructure further enhances its attractiveness as a sustainable solution. Overall, the advancements in microwave-based waste treatment highlight its potential to revolutionize waste management practices, offering a cleaner, more efficient, and adaptable approach to addressing global waste challenges [26–28].

In conclusion, HPEM technology has the potential to transform a multitude of industries by offering efficient, precise, and environmentally friendly solutions. As research and development continue to advance, the integration of HPEM and mm-wave technologies is poised to drive further innovations and improvements across these sectors, paving the way for future technological breakthroughs and sustain-able practices.

1.4 GlobalEM

GlobalEM is a biennial conference that gathers world-class specialists in HPE, HPEM, intentional electromagnetic interference (IEMI), antennas, high-voltage sources, sensors, and numerical methods. This conference is a pivotal platform for experts to exchange ideas, present research findings, and discuss the latest advancements in the field. The contributions in this book reflect the cutting-edge research and innovations shared at the conference, providing valuable insights for engineers and researchers interested in the trends and developments in producing and radiating HPEM.

The meeting has a rich history behind it. In 1978, the late Dr. Carl Baum orga-nized the first Nuclear Electromagnetic Pulse Meeting, or the NEM, in Albuquerque with support from the Summa Foundation, which Carl Baum had founded many years earlier. This first meeting brought scientists and engineers from the US and Western Europe. The NEM Meeting was later renamed as the High-Power Electromagnetics Meeting or HPEM. When this meeting was held in 1994 in Bordeaux, France, it was renamed EUROEM, and subsequently, the meetings in North America were called AMEREM. These meetings have been held every year since 1978. In 2015, due to the increased number of papers from Asia, it was decided to hold the first ASIAEM Conference in Jeju, South Korea. The following meetings were organized in India and China and were called ASIAEM. In 2022, the meeting moved to Abu Dhabi and was renamed GLOBALEM, regardless of the conference location.

1.5 Organization of the book

This book is organized as follows:

- Chapter 2: Semi-Analytical Gray-Box Modeling of Antennas in the Time Domain and Application to Impulse Radiating Antennas. This chapter explores a method to generalize semi-analytical expressions for fields radiated by impulse radiating antennas using a gray-box modeling approach, combining experimental data with known physical behaviors.

- Chapter 3: A Demonstrator for Remote Induction of Disturbance for Access Denial in L-band (RIDAD). This chapter presents the design and integration of a high-power electromagnetic radiator operating in the L-band, intended to test the effects of high-power electromagnetic fields on electronic equipment.
- Chapter 4: Directed Energy Center at the University of New Mexico (DEC@UNM). This chapter introduces the DEC@UNM, a unique academic center performing state-of-the-art research in directed energy, including lasers and microwaves.
- Chapter 5: Compact and Efficient Mode Converter for HPEM Applications in L Band. This chapter discusses the design and integration of a mode converter for a pulsed megawatt source in the L-band, focusing on maximizing power transfer and system efficiency.
- Chapter 6: Design and Characterization of the Tapered Impedance Half Impulse Radiating Antenna, TI-HIRA. This chapter presents experimental results from characterizing a tapered impedance half impulse radiating antenna, highlighting its design and measured performance.
- Chapter 7: Cathode Edge Effect and Divergence of Emitted Electron Beams in Vircators. This chapter studies the electron beam divergence in virtual cathode oscillators (Vircators), aiming to improve their performance and energy efficiency.
- Chapter 8: Use of C-UAS System and Its EM Effect Analysis. This chapter examines the electromagnetic characteristics of counter-UAS (C-UAS) systems and their impact on critical systems, providing insights into their detection and neutralization capabilities.
- Chapter 9: Real-Time Shielding Compromise and Detection. This chapter focuses on detecting and responding to HPEM events, emphasizing the importance of maintaining effective shielding over the life of the system.
- Chapter 10: Financial Comparative Analysis of Substation HPEM Mitigation Designs. This chapter details the design and implementation of a cost-effective solution for HPEM mitigation in electric substations, including real-time detection and monitoring.
- Chapter 11: Three-Dimensional FDTD-based Lightning Transient Analysis of Secondary Circuits With Shielded Control Cables Over Grounding Structures in a Substation. This chapter discusses the use of FDTD simulations to analyze electromagnetic transients in substations, aiming to improve the design of shielded control cables.
- Chapter 12: An Introduction to a Resilience-based Approach to Transient HPEM Disturbance Mitigation. This chapter introduces a resilience-based approach to HPEM threat mitigation, emphasizing the importance of managing HPEM resilience through international standards.
- Chapter 13: Susceptibility of Switch-Mode Power Supplies Due to Conducted Pulse From HEMP. This chapter explores the vulnerability of switch-mode

power supplies to electromagnetic pulses from high-altitude nuclear explosions, providing insights into their susceptibility and mitigation strategies.
- Chapter 14: Electromagnetic Security: Threat Models and Exploitation Scenarios Exploiting Susceptibility to HPEM. This chapter examines the relationship between electromagnetic compatibility and information security, focusing on threats posed by electromagnetic emissions and susceptibility.

The chapters in this book collectively offer a comprehensive view of the latest research and advancements in high-power electromagnetics. By presenting detailed studies and innovative solutions, this book aims to serve as a valuable resource for engineers, researchers, and specialists involved in producing radiation and protecting against unwanted electromagnetic signals. We hope this book will inspire further research and development in this dynamic and critical field.

Acknowledgments

We would like to express our sincere thanks to the Suma Foundation, for organizing GlobalEM and fostering scientific advancement in the area of high-power electromagnetic fields.

We extend our thanks to the authors for their dedication and effort in writing the chapters of this book. Their valuable contributions have been essential to the completion of this work, enriching the knowledge in each of the areas covered.

We also extend our gratitude to the reviewers for their meticulous work in ensuring the quality and scientific rigor of this book. We are thankful to the institutions, universities, and funding agencies that supported the research and enabled the authors to participate in this project.

References

[1]   F. Sabath, "Overview of Four European High-Power Microwave Narrow-Band Test Facilities," *IEEE Transactions on Electromagnetic Compatibility*, vol. 46, no. 3, pp. 329–334, 2004, doi:10.1109/TEMC.2004.831822.

[2]   J. Benford, J. A. Swegle, and E. Schamiloglu, *High Power Microwaves*, 3rd ed. Boca Raton, FL: CRC Press, 2016.

[3]   W. D. Prather, C. E. Baum, J. M. Lehr, *et al.*, "Ultra-Wideband Source and Antenna Research," *IEEE Transactions on Plasma Science*, vol. 28, no. 5, pp. 1624–1632, 2000.

[4]   Ariztia, L., Ibrahimi, N., Zhabin, A., *et al.*, "A High-Power Electromagnetic Source for Disabling Improvised Explosive Devices," *High Voltage*, vol. 9, no. 2, pp. 403–409, 2024. https://doi.org/10.1049/hve2.12416

[5]   K. Yu. Sakharov, A. V. Sukhov, V. L. Ugolev, and Yu. M. Gurevich, "Study of UWB Electromagnetic Pulse Impact on Commercial Unmanned Aerial Vehicle," *Laboratory of EMP Generation and Measurements*, All-Russian Research Institute for Optical and Physical Measurements, Moscow, Russia.

[6]   D. V. Giri, *High-Power Electromagnetic Radiators: Nonlethal Weapons and Other Applications*. Cambridge, MA: Harvard University Press, 2004.

[7]   J. Zhang, D. Zhang, Y. Fan, *et al.* "Progress in Narrowband High-Power Microwave Sources," *Physics of Plasmas*, vol. 27, no. 1, p. 010501, 2020.

[8]   "What Is Drone Jamming and How Can It Be Countered?", Doodle Labs. [Online]. Available: https://doodlelabs.com/what-is-drone-jamming/. [Accessed: Aug. 07, 2024].

[9]   "How Does Jamming Work? Exploring the Tech of Signal Interference," Cyber Insight. [Online]. Available: https://cyberinsight.co/how-does-jamming-work/. [Accessed: Aug. 07, 2024].

[10] ITOPP, "Neutralization of Autonomous and Swarm of Drones," ALCEN, 2023. [Online]. Available: https://www.itopp-alcen.com/sites/itopp-alcen/files/ithpp/pdf/Neutralization-of-Autonomous-and-Swarm-of-Drones-%2001-ITOPP-2023-Web_0.pdf. [Accessed: Aug. 07, 2024].

[11] "Electronic Warfare," Epirus Inc. [Online]. Available: https://www.epirusinc.com/electronic-warfare. [Accessed: Aug. 07, 2024].

[12] S. Park, H. T. Kim, S. Lee, H. Joo and H. Kim, "Survey on Anti-Drone Systems: Components, Designs, and Challenges," *IEEE Access*, vol. 9, pp. 42635–42659, 2021, doi:10.1109/ACCESS.2021.3065926.

[13] B. H. Lipshutz, "The Synthesis of Active Pharmaceutical Ingredients (APIS) Using Continuous Flow Chemistry," *Beilstein Journal of Organic Chemistry*, vol. 11, pp. 1194–1219, 2015, [Accessed: 01-Aug-2024]. [Online]. Available:https://www.beilstein-journals.org/bjoc/articles/11/134

[14] S. Virlley, S. Shukla, S. Arora, *et al.*, "Recent Advances in Microwave-Assisted Nanocarrier Based Drug Delivery System: Trends and Technologies," *Journal of Drug Delivery Science and Technology*, vol. 87, p. 104842, 2023.

[15] M. Selmi, A. A. B. Dukhyil, and H. Belmabrouk, "Numerical Analysis of Human Cancer Therapy Using Microwave Ablation," *Applied Sciences*, vol. 10, no. 1, p. 211, 2020. [Online]. Available: https://www.mdpi.com/2076-3417/10/1/211.

[16] W. M.-S. C. J. Simon, and D. E. Dupuy, "Microwave Ablation: Principles and Applications," *Radiographics*, vol. 25, no. Suppl 1, pp. S69–S83, 2005. [Online]. Available: http://pubs.rsna.org/doi/full/10.1148/rg.25si055501

[17] B. Feng, P. Liang, Z. Cheng, *et al.*, "Ultrasound-Guided Percutaneous Microwave Ablation of Benign Thyroid Nodules: Experimental and Clinical Studies," *European Journal of Endocrinology*, vol. 166, no. 6, pp. 1031–1037, 2012. [Online]. Available: https: //eje.bioscientifica.com/view/journals/eje/166/6/1031.xml

[18] S. Eskandari, S. M. Jalalalhosseini, and E. Mortezazadeh, "Microwave Heating as an Enhanced Oil Recovery Method—Potentials and Effective Parameters," *Energy Sources, Part A: Recovery, Utilization, and Environmental Effects*, vol. 37, no. 7, pp. 742–749, 2015, [Online]. Available: https://doi.org/10.1080/15567036.2011.592906

[19]   Y. Zhang, P. Chen, S. Liu, *et al.*, "Effects of Feedstock Characteristics on Microwave-Assisted Pyrolysis – A Review," *Bioresource Technology*, vol. 246, pp. 307–323, 2017, [Online]. Available: https://doi.org/10.1016/j.biortech.2017.06.098

[20]   X. Tao, N. Xu, M. Xie, and L. Tang, "Progress of the Technique of Coal Microwave Desulfurization," *International Journal of Coal Science & Technology*, vol. 1, pp. 113–128, 2014, [Online]. Available: https://doi.org/10.1007/s40789-014-0013-6

[21]   W. I. Lee and G. S. Springer, "Microwave Curing of Composites," *Journal of Composite Materials*, vol. 18, no. 4, pp. 387–409, 1984, [Online]. Available: https://doi.org/10.1177/002199838401800405

[22]   G. R. Askari, Z. Emam-Djomeh, and S. M. Mousavi, "Effects of Combined Coating and Microwave Assisted Hot-air Drying on the Texture, Microstructure and Rehydration Characteristics of Apple Slices," *Food Science and Technology International*, vol. 12, no. 1, pp. 39–46, 2006, [Online]. Available: https://doi.org/10.1177/1082013206062480

[23]   K. Oglesby, P. Woskov, H. Einstein, and B. Livesay, "Deep Geothermal Drilling Using Millimeter Wave Technology (Final Technical Research Report)," *Technical Report*, 2014. DOI: https://doi.org/10.2172/1169951. OSTI ID: 1169951.

[24]   M. J. Troughton, *Handbook of Plastics Joining: A Practical Guide*, 2nd ed. Norwich, NY: William Andrew Publishing, 2008.

[25]   Y. Nigam, A. Gupta, R. S. Chauhan, *et al.*, "Microwave Welding of Metals and Non-metals (Thermo-Plastics) [Joining or Welding of Two Materials Using Microwave]," in *Proceedings of the International Conference on Recent Advances in Materials, Mechanical and Civil Engineering*, 2021, pp. 1–8.

[26]   H. Xia, C. Li, G. Yang, *et al.*, "A Review of Microwave-Assisted Advanced Oxidation Processes for Wastewater Treatment," *Chemical Engineering Journal*, vol. 335, pp. 896–908, 2018, [Online]. Available: https://doi.org/10.1016/j.cej.2017.11.120

[27]   K. Zimmermann, "Microwave as an Emerging Technology for the Treatment of Biohazardous Waste: A Mini-review," *Waste Management & Research*, vol. 35, no. 5, pp. 471–483, 2017, [Online]. Available: https://doi.org/10.1177/0734242X16684385

[28]   J. Li, J. Tao, B. Yan, L. Jiao, G. Chen, and J. Hu, "Review of Microwave-Based Treatments of Biomass Gasification Tar," *Renewable and Sustainable Energy Reviews*, vol. 150, no. 111510, 2021, [Online]. Available: https://doi.org/10.1016/j.rser.2021.111510

Chapter 2

Semi-analytical gray-box modeling of antennas in the time domain and application to impulse radiating antennas

Elias Le Boudec[1], Nicolas Mora[2], Farhad Rachidi[1], Marcos Rubinstein[3] and Felix Vega[4]

Impulse radiating antennas, proposed by C. E. Baum in 1989 [1], allow to generate impulse-like, broadband, high-amplitude electromagnetic fields. Applications include electromagnetic compatibility testing and remote sensing [2,3]. Analytical and semi-analytical expressions for the electric and magnetic fields emitted by such antennas are challenging to obtain because of the broadband nature of the signals and the interactions between fields, reflectors, feeder plates, and matching resistors. Analytical formulas can be obtained by integration of the tangential electric field on the antenna aperture [1], later completed by the inclusion of the prepulse [4] (that is, the spurious radiation of the feeder plates) and the diffraction of the feeder plates [5]. An analytical formula for the near- and far-fields valid in the boresight cone was presented in [6]. Ad hoc formulas valid everywhere have been developed, notably for impulse-radiating antennas fed by transverse-electromagnetic coplanar plates [7]. However, analytical models that both incorporate the prepulse and are valid outside the aperture of the antenna have not yet been developed.

In this chapter, we propose to generalize such semi-analytical expressions for the fields radiated by any impulse radiating antenna to the whole space thanks to a "gray-box" modeling approach, i.e., we fit experimental data to a model whose physical behavior is known, but specific parameters are not. We first introduce the method, based on the time-domain Cartesian multipole expansion, with some adaptations to inhomogeneous media. Then, we show how to implement the proposed method as an optimization problem. This approach can be seen as a data-driven model calibration.

[1]Electromagnetic Compatibility Laboratory, EPFL (Swiss Federal Institute of Technology in Lausanne), Lausanne, Switzerland
[2]Department of Electrical and Electronics Engineering, Faculty of Engineering, Universidad Nacional de Colombia, Bogota, Colombia
[3]Institute for Information and Communication Technologies, University of Applied Sciences and Arts Western Switzerland, Yverdon-les-Bains, Switzerland
[4]Directed Energy Research Center, Technology Innovation Institute, Abu Dhabi, United Arab Emirates

Next, we apply the technique to an impulse-radiating antenna modeled by a time-domain finite-element simulation and evaluate its performance. Finally, we conclude with some comments.

2.1   Method

In this section, we present the direct model, which maps the parameters (defined in Section 2.1.3) to the predicted fields. Next, we introduce the inverse problem, which seeks optimal parameters and moments given measured fields.

2.1.1   *The time-domain Cartesian multipole expansion*

In the frequency domain and lossless media, the propagation of electromagnetic waves is described by the Helmholtz equation. To solve this equation, one could resort to spherical coordinates and use the separation of variables. Doing this, we obtain the frequency-domain spherical multipole expansion [8], which projects a general solution onto a series of angle-dependent functions (spherical harmonics, the restriction to the sphere of an orthogonal basis of harmonic and homogeneous polynomials in $\mathbb{R}^3$), and radius-dependent functions (spherical Bessel functions of the first or second type, depending on boundary conditions). This method is commonly used in electromagnetic problems; see, for example, [9,10]. In the time domain, however, the Helmholtz equation becomes the wave equation, and such a multipole expansion, derived in spherical coordinates in [11], is less common. In Cartesian coordinates, the multipole expansion offers approximate (semi-analytical) solutions to partial differential equations with constant coefficients through the truncated spherical expansion of the appropriate Green's function [12]. As the truncation order goes to infinity, the multipole expansion converges (in an appropriate sense, e.g., [13]) to the true field under moderate assumptions. This truncated series has a closed-form expression for electromagnetic fields in homogeneous and isotropic media. Indeed, in this case, the Green's function tensor consists of derivatives of $\delta(t - r/c)/(4\pi r)$, where δ is the one-dimensional Dirac δ (Schwartz) distribution. While this analysis is often done in the frequency domain, it is also possible in the time domain [11]. However, one must consider subtleties such as the disappearance of clearly identifiable wavelengths.

Because of its straightforward implementation, we work with the time-domain Cartesian multipole expansion. The general theoretical background of the Cartesian multipole expansion is presented below. Further insight can be found in [11,14,15]. We start from Maxwell's equations in a lossless, homogeneous, and isotropic medium, with sources possessing compact support, to obtain the electric field wave equation:

$$\Box \mathbf{E}(t, \mathbf{r}) = -\frac{1}{\varepsilon}\nabla\rho(t, \mathbf{r}) - \mu\frac{\partial \mathbf{J}(t, \mathbf{r})}{\partial t} \tag{2.1}$$

$$0 = \frac{\partial \rho(t, \mathbf{r})}{\partial t} + \nabla \cdot \mathbf{J}(t, \mathbf{r}) \tag{2.2}$$

where $\Box = \mu\varepsilon(\partial/\partial_t)^2 - \nabla^2$ is the d'Alembert operator, ε is the electrical permittivity and μ the magnetic permeability. Next, we perform in Appendix A the multipole expansion by writing the component $u \in \{x, y, z\}$ of the right-hand side of (2.1) as a generic point source

$$\sum_{|\alpha| \leq n} \frac{(-1)^{|\alpha|}}{\alpha!} \left[C_\alpha^{\frac{\partial \rho}{\partial u}}(t) + C_\alpha^{J_u}(t) \right] \frac{\partial^{|\alpha|} \delta(\mathbf{r} - \mathbf{r}_0)}{\partial x^{\alpha_x} \partial y^{\alpha_y} \partial z^{\alpha_z}} \tag{2.3}$$

where n is the truncation order, $\alpha = (\alpha_x, \alpha_y, \alpha_z) \in \mathbb{N}^3$ is a multi-index, $|\alpha| = \alpha_x + \alpha_y + \alpha_z$, $\alpha! = \alpha_x! \alpha_y! \alpha_z!$, δ is the three-dimensional Dirac δ distribution, $\mathbf{r}_0$ is the position of the point source, the current and charge moments are given by

$$C_\alpha^{\frac{\partial \rho}{\partial u}}(t) = -\frac{1}{\varepsilon} \iiint \frac{\partial \rho}{\partial u}(t, \mathbf{r}) \mathbf{r}^\alpha \mathrm{d}^3 \mathbf{r} \tag{2.4}$$

$$C_\alpha^{J_u}(t) = -\mu \frac{\partial}{\partial t} \iiint J_u(t, \mathbf{r}) \mathbf{r}^\alpha \mathrm{d}^3 \mathbf{r} \tag{2.5}$$

and $\mathbf{r}^\alpha = x^{\alpha_x} y^{\alpha_y} z^{\alpha_z}$. Such an equivalent point source appears as a consequence of performing a Taylor series of Green's function [14] and can represent current densities that are not punctual. Note that there is a correspondence between the spherical multipole expansion and its Cartesian counterpart [12]. Because of the conservation of charge and under the compact-support assumption, the charge moments can be computed from the current moments. A detailed description of the computation of the charge moments from the current moments is presented in Appendix B. Because of the differential equation linking the current and the charge moments, if $\alpha_{\max}$ is the multi-index of the highest-order current moment, then the truncation order must satisfy

$$n \geq |\alpha_{\max}| + 2 \tag{2.6}$$

Indeed, a given current moment will induce a charged moment that is two orders higher.

Observing the current-moment definition (Equation (2.5)), it appears that the time-dependence of each moment is *a priori* allowed to vary independently from the others. As this gives rise to a vast number of degrees of freedom, we simplify by assuming that the time- and space-dependence of the current density are separable, i.e.,

$$\mathbf{J}(t, \mathbf{r}) = h(t)\mathbf{j}(\mathbf{r}) \tag{2.7}$$

Thus, the moments are entirely defined by the time-dependent excitation h and the scalar multipole moments

$$C_\alpha^{j_u} = -\mu \iiint j_u(\mathbf{r}) \mathbf{r}^\alpha \mathrm{d}^3 \mathbf{r} \tag{2.8}$$

As we can only consider finite multipole expansions, the truncation order n determines the number of scalar multipole moments we must consider, i.e., all those whose order $|\alpha| = \alpha_x + \alpha_y + \alpha_z$ is less than or equal to n. The number of such multipole moments grows as $\mathcal{O}(n^3)$. As highlighted in [8, Problem 4.3] for the scalar potential, this cubic growth does not contradict the quadratic growth of spherical harmonics.

Finally, we must compute the field resulting from a generic point source as in (2.3). By linearity of the problem, it is sufficient to consider a single term, given by some multi-index α and its corresponding time-varying multipole moment C_α:

$$C_\alpha(t)\frac{\partial^{|\alpha|}\delta(\mathbf{r})}{\partial x^{\alpha_x}\partial y^{\alpha_y}\partial z^{\alpha_z}} \tag{2.9}$$

In Appendix C, we verify that such a generic time-varying point source radiates the causal field

$$g_\alpha(t - r/c, \mathbf{r}; C_\alpha) \tag{2.10}$$

where r is the radius, $c = 1/\sqrt{\mu\varepsilon}$ is the wave speed, and g_α is an auxiliary function, recursively defined by

$$g_\alpha(t, \mathbf{r}; C_\alpha) = \begin{cases} C_\alpha(t)/(4\pi r) & \alpha = (0,0,0) \\ \dfrac{\partial g_{\alpha-\tilde{\mathbf{e}}_u}(t, \mathbf{r}; C_\alpha)}{\partial u} - \dfrac{\partial g_{\alpha-\tilde{\mathbf{e}}_u}(t, \mathbf{r}; C_\alpha)}{\partial t}\dfrac{u}{rc} & \text{otherwise} \end{cases} \tag{2.11}$$

where $\tilde{\mathbf{e}}_u$ is the unit vector in Cartesian coordinates corresponding to $u \in \{x, y, z\}$, and u is the first coordinate such that $\alpha_u > 0$.

Finally, note that the same procedure applies to the magnetic field. This field obeys a differential equation similar to (2.1) but with the curl of the current density on the right-hand side. The difference in the method is the computation of moments.

2.1.2   Moving to inhomogeneous media

The multipole expansion introduced above assumes the medium to be homogeneous. This hypothesis often fails to hold, e.g., when considering metallic structures such as reflectors or waveguides. However, this limitation can be mitigated with the help of image theory.

Seen as boundary conditions, perfect electric conductor planes set the tangential component of the electric field to zero. Without loss of generality, let us assume that there is such a perfect electric conductor plane on the Oyz-plane. Then, the field radiated by a current density $\mathbf{J}$ restricted to the positive-x half-space can be obtained by removing the perfect electric conductor and extending the current density by including its negative-x half-space image:

$$\mathbf{J}(x < 0, y, z) = \begin{bmatrix} J_x(-x, y, z) \\ -J_y(-x, y, z) \\ -J_z(-x, y, z) \end{bmatrix} \tag{2.12}$$

Thus, the presence of this perfect electric conductor plane can be included in the homogeneous-space multipole expansion by ensuring that all the moments $C_\alpha^{j_u}$ satisfy the corresponding current-density (anti-)symmetry. From (2.8), a necessary condition is that

$$\begin{cases} C_\alpha^{j_y} = C_\alpha^{j_z} = 0 & \text{if } \alpha_x \text{ is even} \\ \quad C_\alpha^{j_x} = 0 & \text{if } \alpha_x \text{ is odd} \end{cases} \tag{2.13}$$

If these conditions are satisfied, the y- and z-components of the electric field vanish on the Oyz plane. Likewise, a perfectly conducting ground (Oxy) plane translates to

$$\begin{cases} C_\alpha^{j_x} = C_\alpha^{j_y} = 0 & \text{if } \alpha_z \text{ is even} \\ \quad C_\alpha^{j_z} = 0 & \text{if } \alpha_z \text{ is odd} \end{cases} \tag{2.14}$$

Similarly, if the electric field is symmetric across the Oxz plane, it is appropriate to include a perfect magnetic conductor by setting the normal component of the current density to zero. This symmetry yields the moment constraint

$$\begin{cases} \quad C_\alpha^{j_y} = 0 & \text{if } \alpha_y \text{ is even} \\ C_\alpha^{j_x} = C_\alpha^{j_z} = 0 & \text{if } \alpha_y \text{ is odd} \end{cases} \tag{2.15}$$

Note that because the current density **J** is a forced input of the model, the source is not affected by the presence of any perfect electric conductor.

As soon as there is more than one reflector (e.g., dihedral planes, rectangular resonators [16,17]), the heuristic introduced in (2.13) no longer holds. Take, for example, a pair of infinite perfect electric conductor planes, say planes A and B. Given an arbitrary localized source **J**, one might start by computing the reflection of **J** through A, $A(\mathbf{J})$. Since $A \circ A$ is the identity (where $\circ$ denotes the composition of functions), the reflection of $A(\mathbf{J})$ through A is **J** itself. However, we now need to reflect both **J** and $A(\mathbf{J})$, yield two new sources $B(\mathbf{J})$ and $B \circ A(\mathbf{J})$. These must each, in turn, undergo all possible reflections, etc.

This process of recursive reflections yields a series of sources. It is reasonable to assume that this series converges to an integrable current density since, at least in practical applications, Joule losses in the conductor limit the efficiency of the reflection. With regards to the multipole expansion, there are at least two possibilities: either these reflections are taken into account before computing the moments (adding constraints such as (2.13)), or the multipole expansion is calculated separately for each original and reflected source. This more general approach goes beyond the scope of this document, which restricts the discussion to simpler symmetries (e.g., perpendicular symmetry planes).

2.1.3 Semi-analytical gray-box modeling

The previous section introduced the direct model, that is, the physical model (the time-domain Cartesian multipole expansion) mapping known inputs (such as the time-dependent current density) to the electric or magnetic fields. Here, we look at the inverse problem, defined as finding inputs that best match measured outputs (electric or magnetic fields). Formally, given measurements (or simulations) of the electric field at the space-time coordinates $(t_i, \mathbf{r}_j)$, the gray-box modeling approach consists of finding the parameter vector **x** (if unique) that minimizes the total prediction error:

$$\text{minimize}_{\mathbf{x} \in \mathbf{R}^{n_{\text{DoF}}}} \sum_{i,j} \| \mathbf{E}(t_i, \mathbf{r}_j) - \tilde{\mathbf{E}}(t_i, \mathbf{r}_j; \mathbf{x}) \|_2^2 \tag{2.16}$$

where n_{DoF} is the number of degrees of freedom (i.e., the dimensionality of the domain), $\mathbf{x}$ is built by stacking the scaled parameters (the sampled time-dependent excitation h, the scalar multipole moments $C_\alpha^{j_u}$, and the equivalent point-source location $\mathbf{r}_0$), $\mathbf{E}(t_i, \mathbf{r}_j)$ is the known electric field, and $\tilde{\mathbf{E}}(t_i, \mathbf{r}_j; \mathbf{x})$ is the corresponding field predicted by a multipole expansion of parameter vector $\mathbf{x}$. Figure 2.1 presents a schematic view of the approach. The entries of the parameter vector $\mathbf{x}$ are unconstrained. Indeed, all predicted fields represent physical solutions, irrespective of the sign or magnitude of entries of $\mathbf{x}$.

This approach corresponds to a semi-transparent (gray-box) model. Indeed, contrary to, e.g., [7], it applies to many time-domain radiation problems, not only impulse radiating antennas. As such, the only *a priori* information available to the model is that its output is a causal electric field radiated by a source satisfying symmetries defined in the section above. Thus, unlike the black-box modeling approaches, which only match the "shape" of data, this approach always delivers physical solutions—i.e., fields satisfying by design Maxwell's equations. On the one hand, this reduces the risk of over-fitting (where irrelevant data features, such as noise, are erroneously learned by the model); equivalently, no explicit regularization is necessary. On the other hand, the physical nature of the model allows us to interpret the output of the minimization physically, such as the equivalent position of the source, its polarity, and the characteristics of the current distribution. Furthermore, it is inherently aligned with the supplied data.

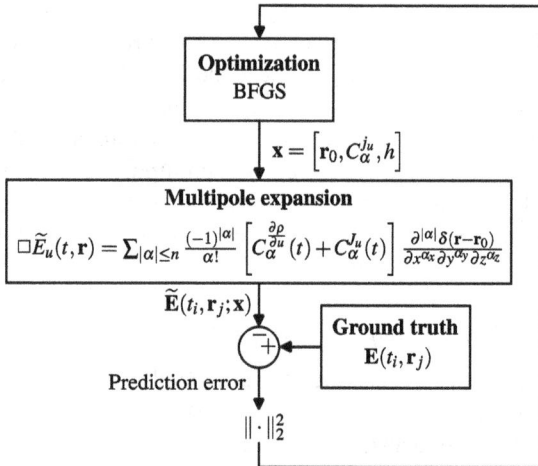

Figure 2.1   Schematic view of the inverse problem, containing the direct problem (i.e., the multipole expansion). The entire loop runs for every iteration of the optimizer (here, BFGS).

2.2   Implementation of the method

This section describes the practical implementation of the direct model and its inverse counterpart. The time-domain multipole expansion is modeled by a custom analytical model, relying on the Python NumPy library for all array-based computations. The radiation of every single multipole moment (dipole, quadrupole, etc.) is pre-computed and stored in a hash map for fast retrieval. For example, the generic term

$$\beta \frac{\mathbf{r}^\alpha}{r^m} \frac{\partial^p h}{\partial t^p} \tag{2.17}$$

with $\beta \in \mathbf{R}$, $\alpha \in \mathbb{N}^3$, and $m, p \in \mathbf{N}$, appearing in the expression of a radiated field, is stored as the hash-map entry

$$(\alpha, m, p) \mapsto \beta \tag{2.18}$$

The total field is then obtained by a linear combination of such terms, weighted by the scalar multipole-moments. The developed implementation is publicly available.[*]

The optimization problem in (2.16) poses several challenges, as it is non-linear and non-convex. Although the gradient of the error with respect to the parameter vector **x** exists, a closed-form expression is difficult to find because of the intricate dependence of the predicted field $\tilde{\mathbf{E}}(t_i, \mathbf{r}_j; \mathbf{x})$ on the model parameters. In turn, we must use zeroth order unconstrained solvers. We use the Broyden–Fletcher–Goldfarb–Shanno (BFGS) algorithm implementation provided by the SciPy library.

As noted above, the discretized time-dependent excitation h is part of the unknown parameters. As its numerical derivatives must also be computed, the required time step is typically tiny, yielding many time samples. In turn, this dramatically increases the number of degrees of freedom, making the convergence of the BFGS algorithm very slow. To solve this, we only look for a small number of points $t_i^\star$ at which h is cubically interpolated. Also, since the measurements $\mathbf{E}(t_i, \mathbf{r}_j)$ span a finite duration in time, the time-dependent excitation h is fixed to the same duration. Finally, as the field depends on derivatives of h, the model's input is defined as the second-order time-derivative h'' to reduce numerical differentiation artifacts.

2.3   Application to an impulse radiating antenna

This section applies the proposed method to an exponentially tapered impulse-radiating antenna.

2.3.1   Data generation through a finite-element simulation

To train the direct model, we use data from a finite-element full-wave simulation implemented thanks to the "electromagnetic waves, transient" COMSOL interface.

---

[*]https://zenodo.org/records/13844777

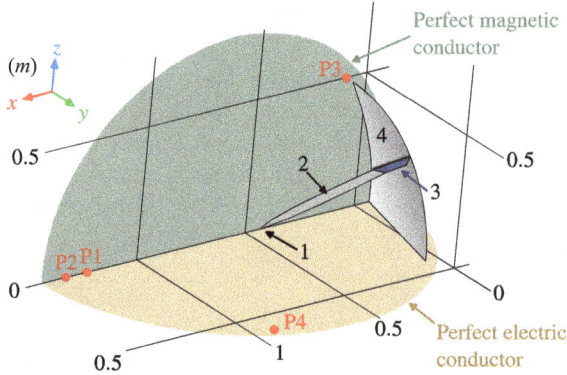

Figure 2.2   Geometry of the COMSOL simulation. The pulse input at the lumped port (1) travels along the feeder plate (2) to reach the matching resistors (3) and then the parabolic reflector (4). The electric field is then sampled at the observation points P1 to P4, whose coordinates are given in Table 2.1.

The geometry, depicted in Figure 2.2, follows the parameters in [18] and uses an exponential feeder profile.

A 50 Ω lumped port ("1" in the figure) located between the ground plane and the transverse-electromagnetic feeder arm (2) transmits a double exponential-like pulse also described in [18]. Note that, in general, the time-dependent excitation h differs from this pulse. Indeed, the former can take into account reflections or dispersion undergone by the latter. Likewise, the equivalent point-source location $\mathbf{r}_0$ is also expected to differ from the lumped port. Next, there are matching resistors (3) between the feeder plates and the parabolic reflector (4). These resistors are modeled as a lossy homogeneous slab whose deeply-subwavelength-thickness and resistivity are such that the lumped resistance corresponds to 200 Ω. A parabolic reflector focuses the incoming spherical wavefront towards the positive x-axis. The lumped port is located at the reflector focal point. The fields are then sampled at all mesh nodes and all time steps. Next, the algorithm is given the electric field samples at the mesh nodes closest to the points given in Table 2.1. Two of these points are placed on the antenna axis. Additionally, two other points are located off the antenna axis—one with a positive y-coordinate and another with a positive z-coordinate—to train the model on off-antenna-axis moments (e.g., y-oriented quadrupole). More points were selected on the antenna axis as it is the area where the fields must fit most precisely.

Ideally, the model would be trained with as many data points as possible to obtain accurate predictions. However, the memory required and the time needed for optimization increase as more data points are added. The conclusion shows that meaningful information about the antenna and extrapolation of the predicted electric field can be extracted using only these four data points. This low number ensures that the optimization process remains tractable on a personal computer.

Finally, to assess the robustness of the proposed method, noise is added to the electric field components. The noise is sampled independently from the standard

Table 2.1   Coordinates of the observation points. The simulated electric field is sampled at the mesh node closest to the above coordinates, and the resulting values are used as input to the inverse problem (see Figure 2.2).

Point number	Coordinate (m)		
	x	y	z
P1	1.3	0	0
P2	1.4	0	0
P3	0.1	0	0.5
P4	0.9	0.6	0

normal distribution with zero mean. Its variance is adjusted for two scenarios: first, by setting the signal-to-noise ratio (SNR) to infinity (i.e., zero noise), and second, by setting the SNR to 0 dB.

2.3.2   Optimization issues

Due to its nonconvexity, the inverse problem applied to an impulse-radiating antenna is a challenging optimization problem. Indeed, there are many local minima of the prediction error—at least the true and zero solutions. In turn, observing the evolution of the prediction error with increasing optimization solver steps, we see plateaux appearing in different configurations, corresponding to local minima. For example, the solver might start moving the equivalent source location far from the origin, momentarily decreasing the prediction error. However, such a distant source is later incapable of satisfyingly predicting the simulated fields. In practice, one can move away from such plateaux by trying several combinations of parameters such as the number of interpolation points $t_i^\star$, the initial parameters vector $\mathbf{x}$, or the scale.

Finally, reducing the number of degrees of freedom to a strict minimum is paramount, which includes restricting the set of possible scalar multipole-moments to those that satisfy the perfect electric and magnetic conductor plane conditions introduced in Eqs. (2.13) to (2.15). We can thus discard what happens in the negative-x half-space (see Figure 2.2), where the fields are negligible because of the reflector. Furthermore, we only sample the electric field in the positive-y half-space. While this causes no issue for the moments because of (2.15), we have to restrict the equivalent source location to the Oxz-plane by setting its y-coordinate to zero.

2.3.3   Results

The equivalent time-dependent excitation h recovered by the optimization algorithm is plotted in Figure 2.3.

The final value of the prediction error in (2.16) drops to 12.7% of the simulated electric field energy for a ninth-order multipole expansion (see Figure 2.6) without noise addition. In other words, the model can explain 87.3% of the energy of the simulated electric field at the observation points. The predicted and simulated fields are shown in Figure 2.4 (no noise addition) and Figure 2.5 (signal-to-noise ratio of 0 dB).

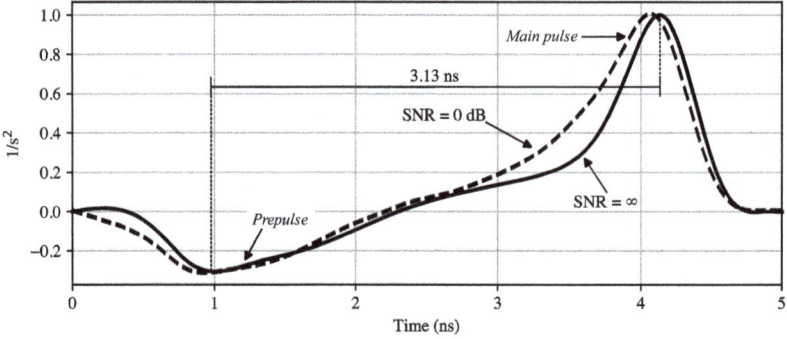

Figure 2.3   Second-order derivative of the equivalent time-dependent excitation h reconstructed by the model. The low-amplitude prepulse is followed by the main pulse, delayed by the time it takes light to cover a round-trip between the port and the reflector (slightly over 3 ns).

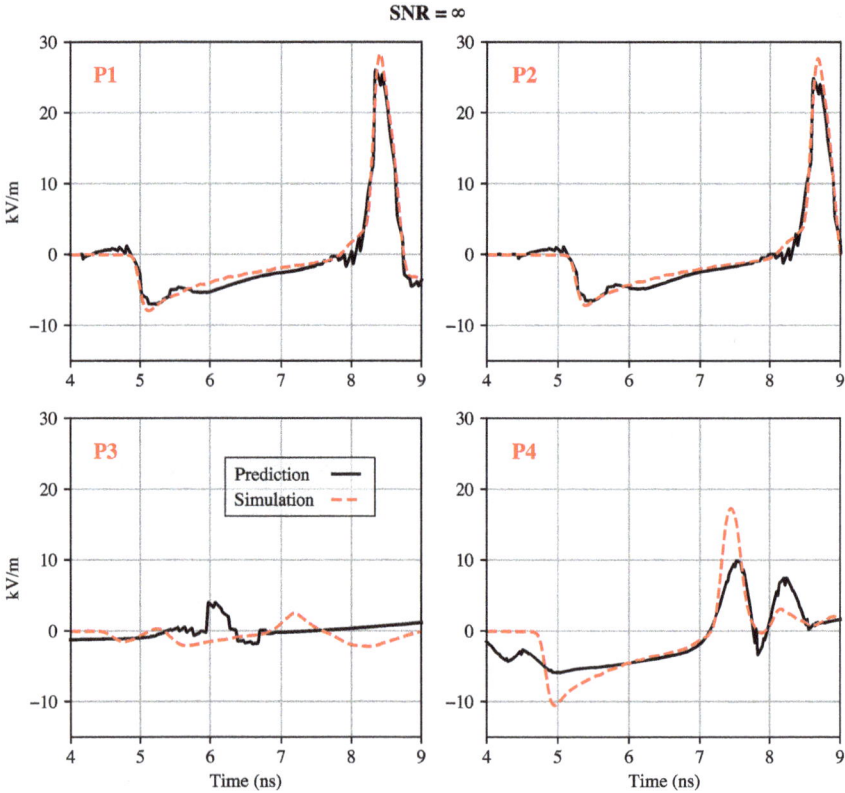

Figure 2.4   Waveforms of the z component of the predicted and simulated electric fields at all observation points (see Figure 2.2) with no noise added to the simulated fields

SNR = 0 dB

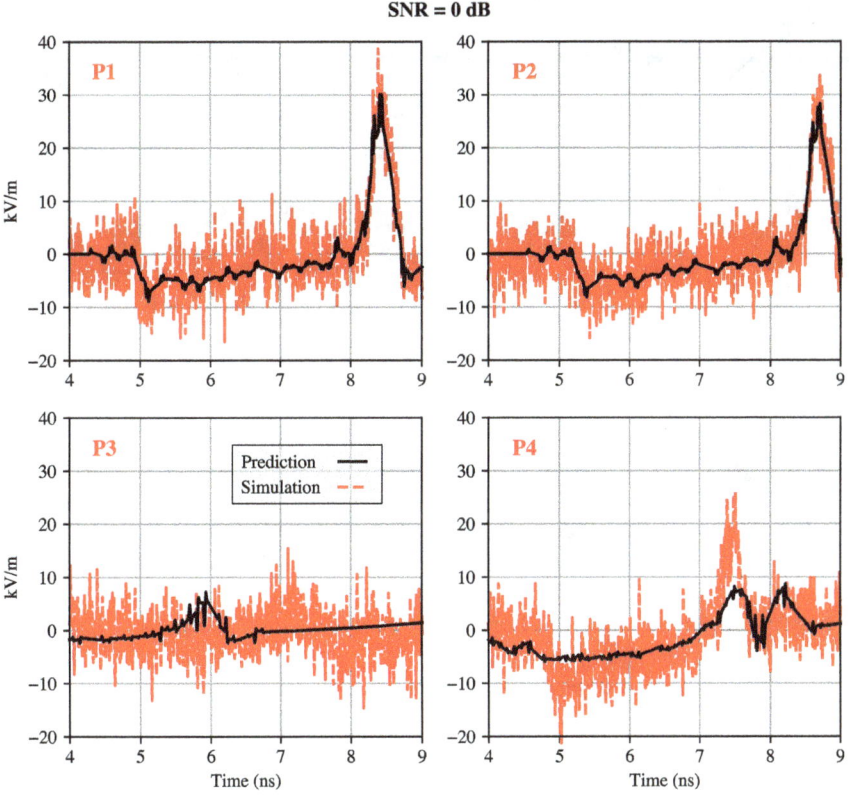

Figure 2.5   Waveforms of the z component of the predicted and simulated electric fields at all observation points (see Figure 2.2) with noise addition (signal-to-noise ratio of 0 dB)

Interpolating the time-dependent excitation h with 45 points $t_i^\star$, the size of the parameter vector $\mathbf{x}$—that is, the number of degrees of freedom n_{DoF}, equals

$$n_{\mathrm{DoF}} = 45 + 2 + |A| \tag{2.19}$$

where $|A|$ is the cardinality of the set A containing all the multi-indices such that $|\alpha| \leq \alpha_{\max}$ and that satisfy Eqs. (2.13) to (2.15), and there is one degree of freedom for the x- and z-coordinates of the equivalent source location $\mathbf{r}_0$. (Recall that $|\alpha_{\max}| = n - 2$, where n is the truncation order.) Figure 2.6 also shows how the number of degrees of freedom evolves for different truncation orders. The equivalent source location is within centimeters of the origin, with no significant variation in the truncation order.

Finally, for a truncation order $n = 5$, the model is dominated by two quadrupoles:

$$C_{(1,0,0)}^{j_z} = -\mu \iiint j_z(\mathbf{r}) x \mathrm{d}^3 \mathbf{r} = 63.3 \, \mathrm{TVsm} \tag{2.20}$$

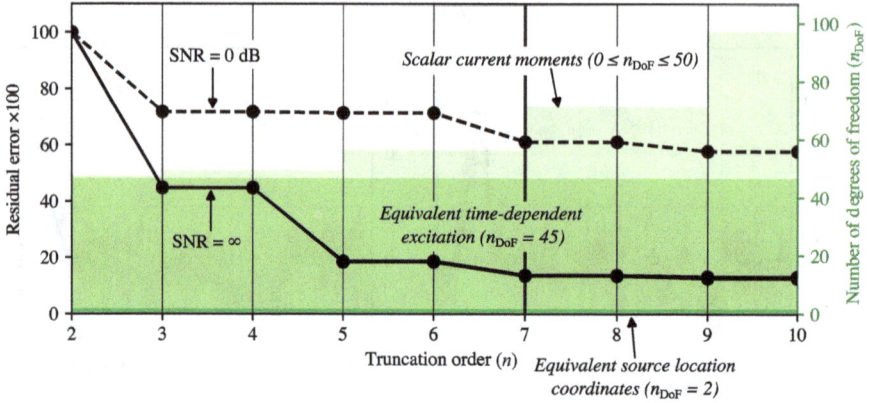

Figure 2.6   Residual error after optimization (normalized to the simulated electric
field energy; solid line: noiseless data, dashed: SNR = 0 dB) and the
number of degrees of freedom (stairs) as a function of the truncation
order n. Admitting up to fifth-order current moments (i.e., the
truncation order is n = 5 + 2, a thicker vertical line in the figure), the
model can predict close to 90% of the energy (Equation 2.16). In that
case, the dimension of the parameter vector **x** is $n_{DoF} = 71$.
Irrespective of n, two degrees of freedom are consumed by the
equivalent source location coordinates (for x and z), 45 for the
interpolation points $t_i^\star$ of the equivalent time-dependent excitation,
and, depending on n, from zero to 50 for the scalar current moments.

$$C_{(0,0,1)}^{j_x} = -\mu \iiint j_x(\mathbf{r})z\,d^3\mathbf{r} = 11.1\,\text{TVsm} \tag{2.21}$$

In turn, the electric field polarity is mostly z-aligned. As a figure of merit, the
contribution of the z-polarized quadrupole on the antenna axis, at $x = 10$ m, corresponds to a peak amplitude

$$E_z^{\text{peak}} \approx \frac{1}{4\pi xc} C_{(1,0,0)}^{j_z} h_{\text{peak}}'' = 1.68\,\text{kVm}^{-1} \tag{2.22}$$

with $h_{\text{peak}}'' = 1$ s^{-2} in Figure 2.3 and $c = 3 \times 10^8$ m s^{-1}. The simulation in [18]
yields a total field amplitude of 1.5 kVm^{-1} at the same distance.

2.4   Discussion and conclusion

In this chapter, we proposed an inverse problem to derive a time-domain semi-
analytical gray-box model of an impulse-radiating antenna. This model is based on
the time-domain Cartesian multipole expansion, usually valid only in homogeneous
media. Here, we showed how this issue could be mitigated thanks to image theory.
The prediction agrees with data from a numerical finite-element simulation for a

simple impulse radiating antenna. It allows us to explain close to 90% of the energy with a relatively low-order multipole expansion. Furthermore, the model can locate an aggregate source close to the reflector. Also, the prepulse and main pulse are visible in the (second-order derivative of the) time-dependent excitation *h*. Finally, the agreement between the model and the simulation is strong in the antenna boresight, which contains most of the energy (in the sense of (2.16)). To further improve the agreement between the shapes of the waveforms outside the antenna boresight, other metrics can be included, such as feature selective validation [19].

The leading moments are two *z*- and *x*-polarized quadrupoles. As the mapping between current density vector fields and multipole moments is not injective, we need additional information to reconstruct a compatible current density. Figure 2.7 shows such a current density. It is given by a field whose space dependence is proportional to

$$\mathbf{r} \mapsto C^{j_z}_{(1,0,0)}\delta(\mathbf{r} - x_0\tilde{\mathbf{e}}_x)\tilde{\mathbf{e}}_z + C^{j_x}_{(0,0,1)}\delta(\mathbf{r} - z_0\tilde{\mathbf{e}}_z)\tilde{\mathbf{e}}_x \qquad (2.23)$$

where $x_0, z_0 > 0$ are the coordinates of the sources. To be consistent with the point source approximation and the separability assumption in (2.7), x_0 and z_0 must be (arbitrarily) close to the origin.

Because of both perfect electric conductor planes, it is easily verified that the proposed current density maps mainly to the appropriate quadrupole moments. Moreover, since the dipoles are located precisely on the perfect electric conductor

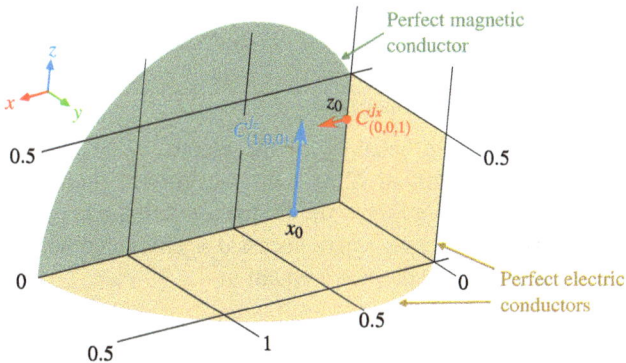

Figure 2.7 A current density vector field yielding the two main quadrupole moments present in the solution for truncation orders n ≥ 3. Two perpendicular dipole sources give this field, each perpendicular to—and located on—the nearest perfect electric conductor plane. The location of the sources on the planes alleviates the problem of recursive reflections discussed earlier in the chapter. The length of the arrow is proportional to the corresponding quadrupole moment amplitude, and the circle at the base of the arrow indicates the source position.

planes, the issue of recursive reflections presented above is absent. Indeed, since the two dipole sources are perpendicular to—and located on—their corresponding perfect electric conductor plane, they are reflected to themselves through this plane. Also, since the planes are perpendicular, only one reflection per dipole source needs to be considered.

As a general comment, compared to traditional methods, both the advantages and disadvantages of the proposed approach lie in its reliance on the data and its "gray-box" modeling approach: indeed, convergence can be slow because of the large parameter space and non-linear problem formulation, but on the other hand, it is specifically designed to adhere to the measurements. This approach could be compared to physics-infused machine learning (e.g., [20]), whose extrapolation capacity is demonstrated in the predicted peak field amplitude in (2.22). Finally, the proposed approach is robust to noise, as is apparent from the residual error in Figure 2.6. Indeed, this error increases with a lower signal-to-noise ratio (SNR), which is expected from the loss of information concomitant with a low SNR. However, the error reduces as the number of degrees of freedom increases. Moreover, the model can extract the time-dependent excitation h from the noisy data (see Figure 2.3). The predicted fields in Figure 2.5 also show a filtering effect. To conclude, possible applications of the proposed method include predicting the field of a given antenna at arbitrary positions while ensuring a good fit of the predictions at known observation points and performance evaluation of an antenna by analyzing the reconstructed multipole moments.

Appendix A

Time-domain multipole moments

This appendix shows how the behavior of the electric field, described in (2.1), yields the multipole expansion in (2.3) with the current and charge moments described by (2.4) and (2.5). We follow the frequency-domain approach in [13] and perform inverse Fourier transforms where necessary. This reference is based on the vector potential, which follows the wave equation[†]

$$\Box \mathbf{A}(t, \mathbf{r}) = \frac{1}{c} \mathbf{J}(t, \mathbf{r}) \tag{A.1}$$

The approach in the cited reference yields the current moments

$$\tilde{C}_\alpha^{J_u}(t) = \iiint J_u(t, \mathbf{r}) \mathbf{r}^\alpha \mathrm{d}^3 \mathbf{r} \tag{A.2}$$

where $u \in \{x, y, z\}$. The multipole expansion then consists in solving the wave equation for the radiating point source

---

[†]In the reference, a 4π factor is also present. This results from a different Green's function convention.

$$\frac{1}{c} \sum_{|a| \leq n} \frac{(-1)^{|a|}}{a!} \tilde{C}_a^{J_u}(t) \frac{\partial^{|a|} \delta(\mathbf{r} - \mathbf{r}_0)}{\partial x^{a_x} \partial y^{a_y} \partial z^{a_z}} \tag{A.3}$$

We proceed by inspection to adapt the definitions above to those introduced in this chapter. It is sufficient to replace the uth component of the current density in (A.1) with the corresponding component of (2.1):

$$\Box E_u(t, \mathbf{r}) = -\frac{1}{\varepsilon} \frac{\partial \rho(t, \mathbf{r})}{\partial u} - \mu \frac{\partial J_u(t, \mathbf{r})}{\partial t} \tag{A.4}$$

The first term corresponds to $C_a^{\frac{\partial \rho}{\partial u}}$ and the second to $C_a^{J_u}$. Their sum, replacing $\frac{1}{c} \mathbf{J}_a$ in (A.3), yields (2.3).

Appendix B

Computation of the charge moments from the current moments

In this appendix, we show how to obtain the scalar charge moments from the scalar current moments, i.e., we look for an expression linking

$$C_a^{\frac{\partial \rho}{\partial u}} = -\frac{1}{\varepsilon} \iiint \mathbf{r}^a \frac{\partial \rho}{\partial u}(\mathbf{r}) \mathrm{d}^3 \mathbf{r} \tag{B.1}$$

where $u \in \{x, y, z\}$, and the scalar current moment, (2.5). ρ denotes the space dependence of the charge density. By the separability assumption in (2.7), we rewrite the conservation of charge (Equation (2.2)) as

$$\nabla \cdot \mathbf{j}(\mathbf{r}) + \rho(\mathbf{r}) = 0 \tag{B.2}$$

Inserting this into the definition of the scalar current moment, we obtain

$$C_a^{\frac{\partial \rho}{\partial u}} = -\frac{1}{\varepsilon} \iiint \mathbf{r}^a \frac{\partial \rho}{\partial u}(\mathbf{r}) \mathrm{d}^3 \mathbf{r} = \frac{1}{\varepsilon} \iiint \mathbf{r}^a \frac{\partial}{\partial u} \nabla \cdot \mathbf{j}(\mathbf{r}) \mathrm{d}^3 \mathbf{r} \tag{B.3}$$

Writing the divergence in Cartesian coordinates, we get

$$C_a^{\frac{\partial \rho}{\partial u}} = \frac{1}{\varepsilon} \sum_{v \in \{x,y,z\}} \iiint \mathbf{r}^a \frac{\partial^2}{\partial u \partial v} j_v(\mathbf{r}) \mathrm{d}^3 \mathbf{r} \tag{B.4}$$

Next, we use integration by parts and the compact support of $\mathbf{j}$:

$$C_a^{\frac{\partial \rho}{\partial u}} = \frac{1}{\varepsilon} \sum_{v \in \{x,y,z\}} \iiint \left[\frac{\partial^2}{\partial u \partial v} \mathbf{r}^a \right] j_v(\mathbf{r}) \mathrm{d}^3 \mathbf{r} \tag{B.5}$$

Now, a direct computation shows that

$$\frac{\partial^2}{\partial u \partial v} \mathbf{r}^{\alpha} = \alpha_v 1_{\alpha_v \geq 1} \cdot \begin{cases} 1_{\alpha_v \geq 2}(\alpha_v - 1)v^{\alpha_v - 2} w_1^{\alpha_{w_1}} w_2^{\alpha_{w_2}} & \text{if } u = v \\ 1_{\alpha_u \geq 1} v^{\alpha_v - 1} \alpha_u u^{\alpha_u - 1} w^{\alpha_w} & \text{otherwise} \end{cases} \tag{B.6}$$

where w_1, w_2, and w are such that

$$\{x, y, z\} = \begin{cases} \{v, w_1, w_2\} & \text{if } u = v \\ \{u, v, w\} & \text{otherwise} \end{cases} \tag{B.7}$$

We get the desired expression since (B.5) is a linear combination of scalar current moments.

Appendix C

Time-domain Cartesian solutions of the wave equation

In this appendix, we show that

$$\Box g_{\alpha}(t - r/c, \mathbf{r}; C_{\alpha}) = C_{\alpha}(t) \frac{\partial^{|\alpha|} \delta(\mathbf{r})}{\partial x^{\alpha_x} \partial y^{\alpha_y} \partial z^{\alpha_z}} \tag{C.1}$$

for all multi-indices α. To proceed, we must show two statements. First, that $\Box g_0(t - r/c, \mathbf{r}; C_0) = C_0(t)\delta(\mathbf{r})$. This is a direct consequence of $-1/(4\pi r)$ being Green's function for the Laplacian ∇^2. Second, we need to show the recursion step, that is, for any $u \in \{x, y, z\}$,

$$\Box g_{\alpha + \tilde{\mathbf{e}}_u}(t - r/c, \mathbf{r}; C_{\alpha + \tilde{\mathbf{e}}_u}) = C_{\alpha + \tilde{\mathbf{e}}_u}(t) \frac{\partial^{|\alpha| + 1} \delta(\mathbf{r})}{\partial x^{\alpha_x} \partial y^{\alpha_y} \partial z^{\alpha_z} \partial u} \tag{C.2}$$

whenever

$$\Box g_{\alpha}(t - r/c, \mathbf{r}; C_{\alpha + \tilde{\mathbf{e}}_u}) = C_{\alpha + \tilde{\mathbf{e}}_u}(t) \frac{\partial^{|\alpha|} \delta(\mathbf{r})}{\partial x^{\alpha_x} \partial y^{\alpha_y} \partial z^{\alpha_z}} \tag{C.3}$$

To this end, note that

$$\frac{\partial g_{\alpha}(t - r/c, \mathbf{r}; C_{\alpha + \tilde{\mathbf{e}}_u})}{\partial u} = \left[\frac{\partial g_{\alpha}}{\partial t}\right](t - r/c, \mathbf{r}; C_{\alpha + \tilde{\mathbf{e}}_u}) \overbrace{\frac{\partial}{\partial u}(t - r/c)}^{-\frac{u}{rc}} + \left[\frac{\partial g_{\alpha}}{\partial u}\right](t - r/c, \mathbf{r}; C_{\alpha + \tilde{\mathbf{e}}_u}) \tag{C.4}$$

thus

$$\Box g_{\alpha + \tilde{\mathbf{e}}_u}(t - r/c, \mathbf{r}; C_{\alpha + \tilde{\mathbf{e}}_u})$$
$$= \Box \left\{ \left[\frac{\partial g_{\alpha}}{\partial u}\right](t - r/c, \mathbf{r}; C_{\alpha + \tilde{\mathbf{e}}_u}) - \frac{u}{rc}\left[\frac{\partial g_{\alpha}}{\partial t}\right](t - r/c, \mathbf{r}; C_{\alpha + \tilde{\mathbf{e}}_u}) \right\} \tag{C.5}$$

by the recursive definition of the auxiliary function. Using (C.4), this becomes

$$\frac{\partial}{\partial u}\Box g_\alpha(t - r/c, \mathbf{r}; C_{\alpha+\tilde{\mathbf{e}}_u}) \tag{C.6}$$

after exchanging the order of derivatives. The right-hand side finally translates to

$$\frac{\partial}{\partial u}C_{\alpha+\tilde{\mathbf{e}}_u}(t)\frac{\partial^{|\alpha|}\delta(\mathbf{r})}{\partial x^{\alpha_x}\partial y^{\alpha_y}\partial z^{\alpha_z}} = C_{\alpha+\tilde{\mathbf{e}}_u}(t)\frac{\partial^{|\alpha|+1}\delta(\mathbf{r})}{\partial x^{\alpha_x}\partial y^{\alpha_y}\partial z^{\alpha_z}\partial u} \tag{C.7}$$

by (C.3), which shows the recursion step.

References

[1] Baum CE. Radiation of impulse-like transient fields; 1989. *Sensor and Simulation Notes*, Note 321.

[2] Baum CE, Baker WL, Prather WD, *et al.* JOLT: a highly directive, very intensive, impulse-like radiator. *Proceedings of the IEEE*. 2004;92(7): 1096–1109.

[3] Shyamala D, Kichouliya R, Kumar P, *et al.* Experimental studies and analysis on IEMI source, field propagation and IEMI coupling to power utility system. *Progress in Electromagnetics Research C*. 2018;83: 229–244.

[4] Farr EG, and Baum CE. Prepulse associated with the TEM feed of an impulse radiating antenna; 1992. Sensor and Simulation Notes, Note 337.

[5] Giri DV, and Baum CE. Reflector IRA design and boresight temporal waveforms; 1994. *Sensor and Simulation Notes*, Note 365.

[6] Mikheev OV, Podosenov SA, Sakharov KY, *et al.* New method for calculating pulse radiation from an antenna with a reflector. *IEEE Transactions on Electromagnetic Compatibility*. 1997;39(1):48–54.

[7] Vega F, and Rachidi F. A simple formula expressing the fields on the aperture of an impulse radiating antenna fed by TEM coplanar plates. *IEEE Transactions on Antennas and Propagation*. 2018;66(3):1549–1552.

[8] Jackson JD. *Classical Electrodynamics*. 3rd ed. New York: Wiley; 1999.

[9] Talashila R, and Ramachandran H. Multipole expansion of radiation from patch antenna using quasi-static surface currents. *IEEE Antennas and Wireless Propagation Letters*. 2020;19(12):2136–2140.

[10] Xia T, Meng LL, Liu QS, *et al.* A low-frequency stable broadband multilevel fast multipole algorithm using plane wave multipole hybridization. *IEEE Transactions on Antennas and Propagation*. 2018;66(11):6137–6145.

[11] Shlivinski A, and Heyman E. Time-domain near-field analysis of short-pulse antennas I. Spherical wave (multipole) expansion. *IEEE Transactions on Antennas and Propagation*. 1999;47(2):271–279.

[12] Rowe EGP. Spherical delta functions and multipole expansions. *Journal of Mathematical Physics*. 1978;19(9):1962–1968.

[13] Wünsche A. Schwache Konvergenz von Multipolentwicklungen. *ZAMM – Journal of Applied Mathematics and Mechanics*. 1975;55(6):301–319.

[14] Kocher CA. Point-multipole expansions for charge and current distributions. *American Journal of Physics*. 1978;46(5):578–579.

[15] Le Boudec E, Kasmi C, Mora N, *et al.* The time-domain Cartesian multipole expansion of electromagnetic fields. *Scientific Reports*. 2024;14:8084.

[16] Harrington RF. *Time-Harmonic Electromagnetics*. New York: Wiley-IEEE Press; 2001.

[17] Tamayo J, López-Peña S, Mattes M, *et al.* A recursive acceleration technique for static potential green's functions of a rectangular cavity combining image and modal series. *IEEE Transactions on Microwave Theory and Techniques*. 2011;59(3):542–551.

[18] Vega F, Albarracin-Vargas F, Kasmi C, *et al.* The tapered impedance half-impulse radiating antenna. *IEEE Transactions on Antennas and Propagation*. 2021;69(2):715–722.

[19] Duffy AP, Martin AJM, Orlandi A, *et al.* Feature selective validation (FSV) for validation of computational electromagnetics (CEM). Part I – The FSV method. *IEEE Transactions on Electromagnetic Compatibility*. 2006;48(3): 449–459.

[20] Iqbal R, Behjat A, Adlakha R, *et al.* Auto-differentiable transfer mapping architecture for physics-infused learning of acoustic field. *IEEE Transactions on Artificial Intelligence*. 2023; p. 1–15.

Chapter 3

A demonstrator for Remote Induction of Disturbance for Access Denial in L-band (RIDAD)

*Adamo Banelli[1], John J. Pantoja[2], Luciano P. de Oliveira[3],
Abdul Baba[3], Mae AlMansoorori[3], Asilah Almesmari[3],
Jesus Alvarez[3], Ernesto Neira[3], Felix Vega[3] and
Chaouki Kasmi[3]*

3.1 Introduction

Critical infrastructure systems in defense, energy, and related fields should be tested against multiple threats, among them, high-power microwave interference and IEMI resilience [1].

Laboratory equipment for emulating these threats requires specialized high-power electromagnetic radiators, capable of producing high-power illuminating fields in controlled conditions.

This chapter presents a demonstrator of a High-Power Electromagnetic Radiator operating in the L band for *Remote Induction of Disturbance for Access Denial* (RIDAD). The system is intended to be a cost-effective narrowband device for testing high-power electromagnetic effects on electronic equipment in laboratory conditions.

The chapter provides detailed information about the design and integration of several of the subsystems used, namely power source, cooling system, mode converter, horn antenna and lens; and other ancillary system.

Measurements conducted in an anechoic chamber show that the system delivered an electric field of 60 kV/m, normalized at 1 m distance, at a frequency of 1.3 GHz.

3.2 System architecture

Figure 3.1 presents the architecture of the RIDAD System. It comprises a 1 MW pulsed magnetron, a high-voltage pulse modulator, a high-voltage peaking unit, a

[1]NSI-MI Technologies, Tuscany, Italy
[2]Single-Photon Group, School of Engineering and Physical Sciences, Heriot-Watt University, Edinburgh, UK
[3]Directed Energy Research Center, Technology Innovation Institute, Abu Dhabi, United Arab Emirates

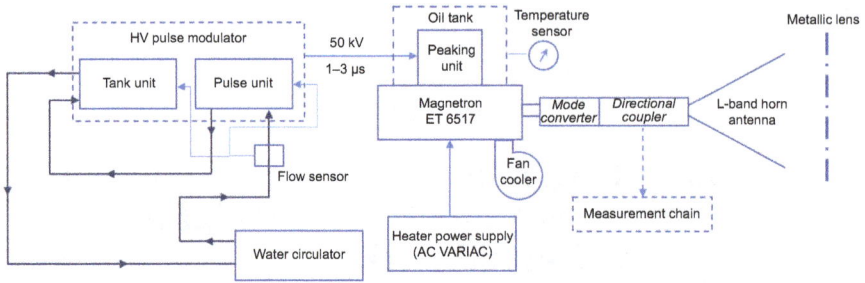

Figure 3.1   Schematic of RIDAD—HPEM radiator prototype in L band

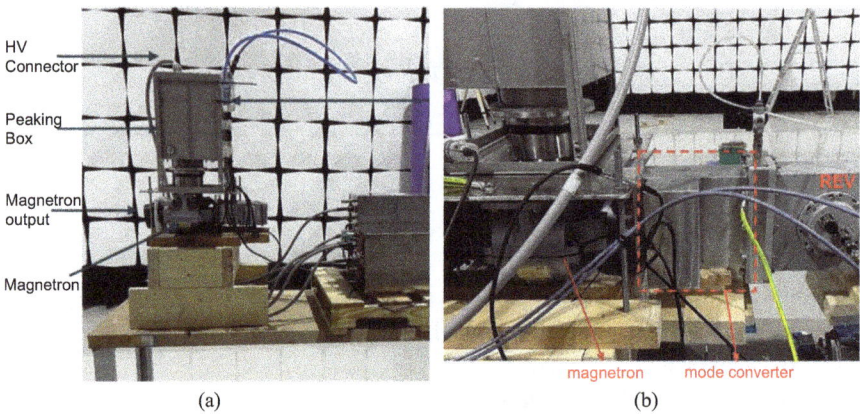

(a)   (b)

Figure 3.2   Transmitter testbed. (a) Cabling check showing the magnetron, high-voltage pulse modulator, and peaking box. (b) Connections between magnetron output, mode converter, and directional coupler.

filament heater unit, a mode converter, and a radiating segment integrating a horn antenna and a metallic lens. Finally, the system includes a set of ancillary systems, such as enclosure, cooling, remote control, and monitoring.

A view of the testing setup is shown in Figure 3.2(a) where the magnetron, the modulator, and the high-voltage peaking box can be seen. Figure 3.2(b) presents connections between the magnetron output, mode converter, and directional coupler.

3.3   Pulsed magnetron

The pulsed magnetron is a Raytheon RK6517 [2,3] and it can be seen in Figure 3.3. The high-voltage connector, frequency tuner, and microwave output are highlighted. The magnetron is an integral magnet type.

The peak power delivered by this unit is nominally 1 MW, although in certain conditions it can be pushed to 1.3 MW during short bursts operation.

The device is mechanically tunable from 1,250 to 1,350 MHz.

Due to the high filament current required, an AC variac, independent from the modulator was used.

Table 3.1 summarizes the general specifications.

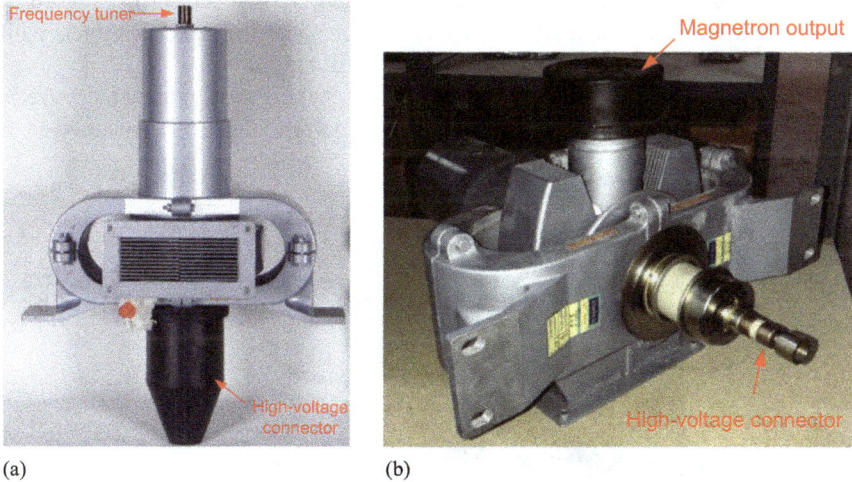

(a)   (b)

Figure 3.3   Magnetron Raytheon RK6517. (a) Side view: high-voltage connector and frequency tuner. (b) Perspective: magnetron output and high-voltage connector.

Table 3.1   General specifications of the pulsed magnetron [3]

Heater current	75 A
Heater voltage at the heater current	26–3.0 V
Minimum heater time	10 min
Cold heater resistance	7.6 m-ohms
Heater current	90 A
Peak anode voltage	70 kV
Peak anode current	60 A
Average input power	4.3 kW
Peak input power	3.5 MW
Max VSWR	1.5/1
Frequency pulling at VSWR 1.5/1	5 MHz
Max pulse duration	3.3 μs
Duty cycle	0.0013
Voltage pulse rise time	0.8 μs
Anode temperature	100 °C
Bushing temperature	150 °C
Output pressurization	45 psi
Pulse repetition rate (PRF)	1–50 Hz

3.4   High-voltage pulse modulator

The high-voltage pulse modulator is Scandinova M-100i [4]. This a solid-state unit comprises a pulse unit that generates and controls the primary voltage pulse and a tank unit responsible for up-converting the pulse to the required high-voltage level.

Figure 3.4 shows the modulator's electric and cooling interfaces. Figure 3.5 presents a detailed block diagram. Table 3.2 presents the specifications, while Figure 3.6 shows a diagnostic waveform of 1 μs duration.

The high-voltage segment also comprises a peaking unit, connected in series with the tank unit. This unit conditions the signal applied to the magnetron by

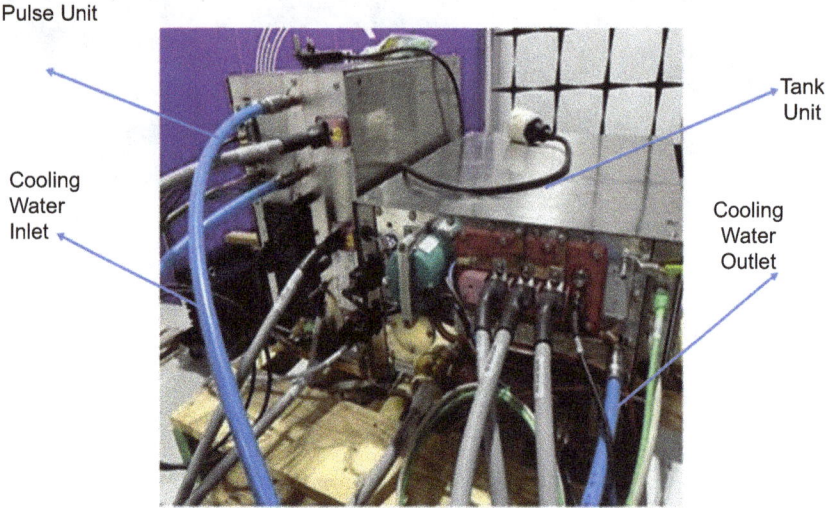

Figure 3.4   Scandinova M-100i high-voltage modulator pulse and tank units

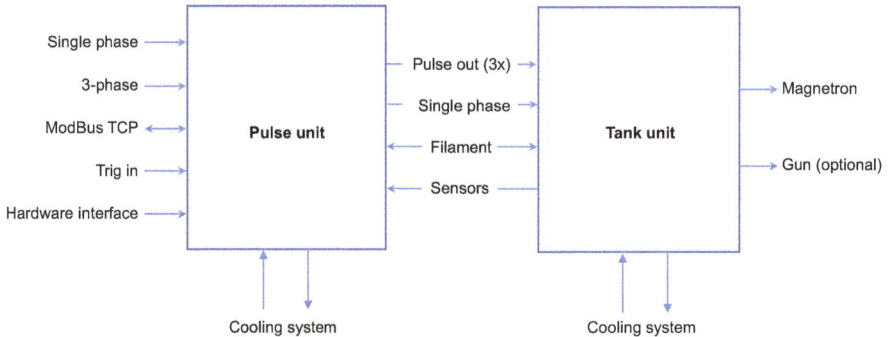

Figure 3.5   Scandinova M-100i modulator block diagram

Table 3.2   Scandinova M-100i modulator general specifications [4]

Specs	Data
Magnetron RF peak power	1–3.1 MW
Magnetron RF average power	2.8 kW
Modulator peak power	6.2 MW
Modulator average power	8 kW
Pulse voltage	30–52 kV
Pulse current	30–120 A
Pulse repetition rate (PRF)	0–500 Hz
RF pulse length	0.5–5 μs
Modulator voltage stability, RMS	0.4%
Cooling	Water
Interface	Default
Mains power, 3 phase	400 VAC, 50/60 Hz
Mains power, single phase	230 VAC, 50/60 Hz
Control interface	ModBus TPC
Water cooling interface in/out	Legris push-in 12 mm
Trig input	Electrical
Diagnostics	Pulse voltage and current

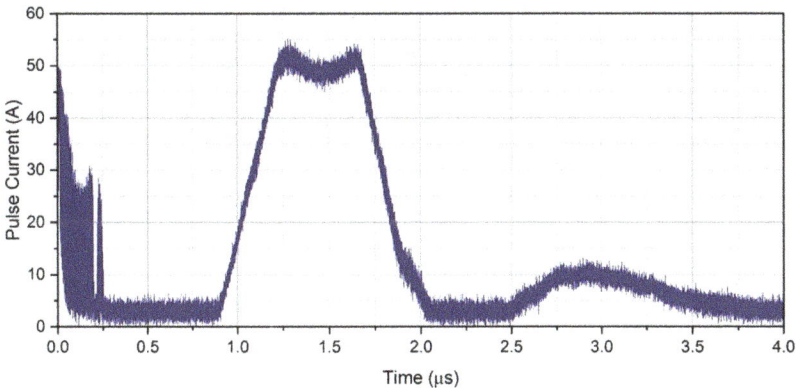

Figure 3.6   Pulse current waveform used for diagnostic during the system

shortening the rise time of the high-voltage waveform produced by the modulator. The peaking unit is presented in Figure 3.7.

3.5   Cooling system

3.5.1   System-level thermal management

The system requires proper cooling to manage heat dissipation within its components, particularly unwanted heat generated by the magnetron, pulse, tank, and

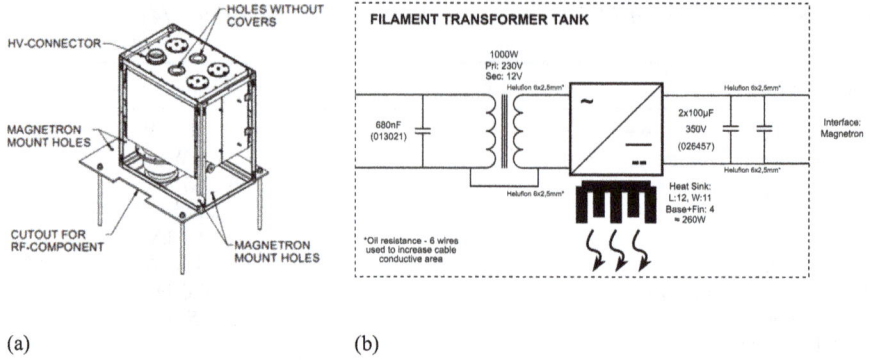

(a)   (b)

Figure 3.7   Peaking box, filament transformer tank. (a) Mechanical schematic.
(b) Electrical schematic.

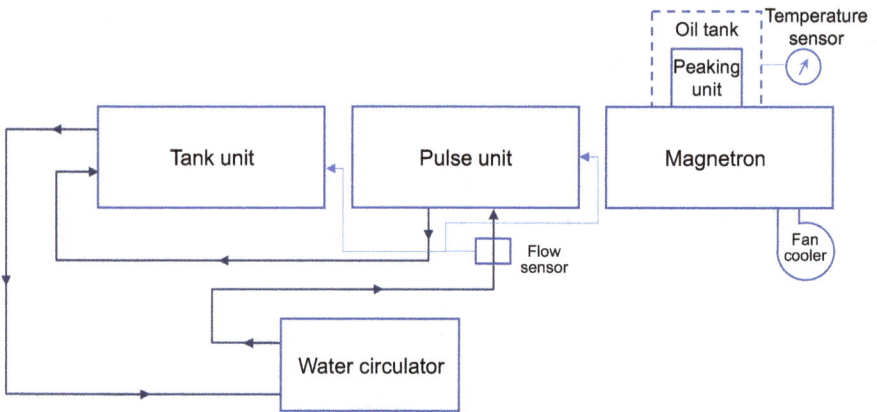

Figure 3.8   Cooling circuit block diagram

peaking units. For appropriate thermal management, two heat removal processes were employed. Figure 3.8 presents the basic cooling block diagram.

3.5.2   Magnetron thermal management

As the magnetron uses forced air-cooling, the ambient temperature will determine the required flow rate to keep the anode temperature below the manufacturer's recommended maximum of 100 °C [2,3].

This was accomplished by forced air cooling with a 200 (c.f.m) axial fan, assuming the magnetron runs at its maximum output power and 25° C ambient temperature conditions.

Figure 3.9 shows the anode temperature rise above the ambient temperature as a function of cooling airflow and different anode power dissipation. It also illustrates the back pressure as a function of airflow rate (c.f.m).

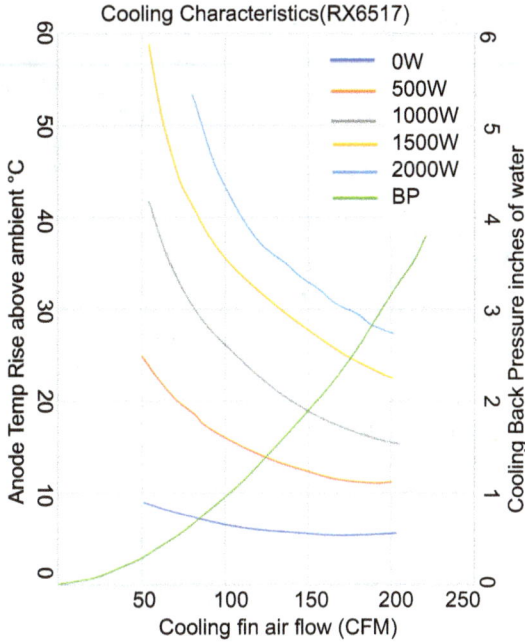

Figure 3.9   Magnetron anode temperature versus cooling airflow for different rates of dissipated power at the anode

3.5.3   Thermal management of the peaking unit

The Peaking Unit was refrigerated and insulated using transformer oil. A 20-liter oil tank reservoir maintained the cathode bushing temperature below the maximum recommended temperature of operation (150° C). The tank also contains the peaking capacitors used to shorten the high-voltage pulse produced by the modulator.

The temperature within the peaking unit was monitored using a K-type thermocouple placed directly on the wall of the metallic oil storage tank, as shown in Figure 3.7. The dimensions of the peaking unit were specified so that a minimum distance of 40 mm was maintained through the oil from high-voltage terminals to the ground with design consideration to avoid sharp objects in the vicinity of the ceramic bushing.

3.5.4   Thermal management of the modulator

The modulator was cooled using an external cooling-water circulator with an adjustable flow rate of 6 to 10 l/min. The operating temperature and fluid flow requirements are specified in Table 3.3. The refrigeration circuits of the tank and pulse units were connected in series.

A flow sensor interlock was installed in the water circuit to prevent the accidental activation of the magnetron without proper refrigeration. Further information regarding flow direction is provided in the block diagram in Figure 3.8.

Table 3.3   Cooling service requirements

Subsystem	Specification	Connection type and cooling capacity requirements
Pulse Unit	• 6 l/min limited by a maximum water pressure of 8 bar • Pressure drop: 0.4 bar • Water temperature: 10 °C–40 °C (non-condensing)	Connected to 12 mm copper tubing requires an external circulator, and the flow direction is specified in Figure 3.8
Tank Unit	• 6 l/min limited by a maximum water pressure of 8 bar • Pressure drop: 0.4 bar • Water temperature: 10 °C–40 °C (non-condensing)	Connected to 12 mm copper tubing requires an external circulator, and the flow direction is specified in Figure 3.8
Filament Transformer Tank (peaking unit)	• Fin type heat sink L 12 cm × W 11 cm Base + fin: 4 cm. • 20-liter transformer oil tank	Cooling of AC DC converter in a peaking unit. Cooling capacity requirement 260 W

(a)   (b)

Figure 3.10   Mode Converter. (a) CAD model. (b) Prototype.

3.6   Mode converter

The magnetron has a circular, non-standard waveguide output, insulated in ceramic. The antenna uses a standard WR650, rectangular waveguide port. Therefore, a mode converter is required for efficient mode conversion and interface conversion, maintaining low return loss [5,6].

Figure 3.10 shows the mode CAD model and the actual realization of the Mode Converter.

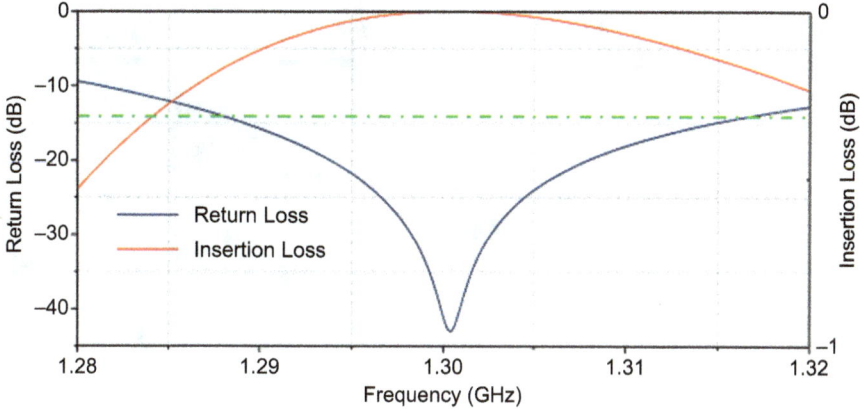

Figure 3.11   Mode converter's return loss and insertion loss

The simulation results presented in Figure 3.11, show that the efficiency of the mode converter is higher than 97% over the frequency range from 1.29 GHz to 1.31 GHz, with a return loss of less than −14 dB over the bandwidth. The conversion efficiency is about 99.3% at the central frequency, with a return loss of approximately −42 dB at 1.30 GHz.

Additional details on the design of the Mode Converter are presented in Chapter 10 of this book.

3.7   Horn antenna and metallic lens

The radiated segment of RIDAD is composed of a standard LB650-15 horn antenna with a gain of 16 dBi and a planar metallic lens added in order to increase the radiated field, without increasing dramatically the overall dimensions of the radiating segment.

Generally, the design of low-profile planar lenses relies on analyzing a unit element with periodic boundary conditions. However, actual lenses have finite elements, and truncating the structure can lead to inaccurate results [7,8]. The Theory of Characteristic Modes (TCM) is a suitable alternative to this issue [9,10].

The planar single-layer lens was designed based on TCM concepts, optimized in CST Design Studio [11] and manufactured in-house. The designed configuration can be seen in Figure 3.12. It consists of an outer ring with a diameter of 1,005 mm and an inner ring with a diameter of 880 mm. Both rings are 22 mm wide and are short-circuited by 48 sectors, as shown in Figure 3.12. The lens is placed 45 cm in the antenna aperture.

To improve the electromagnetic field distribution on the aperture of the antenna, an 86-mm wide metal brim was added, as shown in Figure 3.12(b).

The lens design also considered the maximum VSWR of 1.5 recommended by the magnetron manufacturer. Figure 3.13 compares the simulated VSWR of the

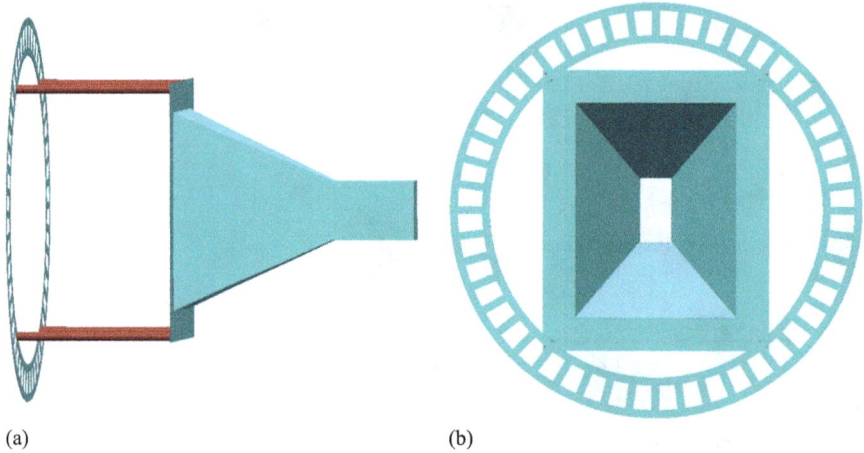

(a)   (b)

Figure 3.12   VSWR comparison—horn and horn antenna and metallic lens

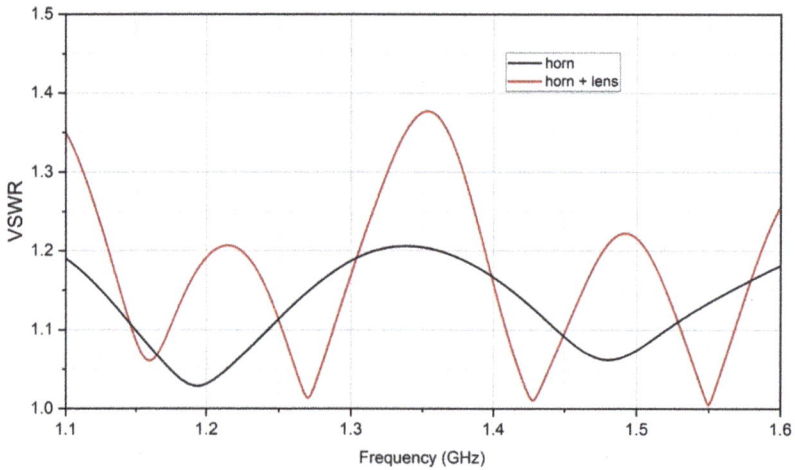

Figure 3.13   Horn antenna and planar metallic lens. (a) Front view. (b) Side view.

single horn antenna and the assembly horn antenna and metallic lens. As can be seen, the VSWR of the final configuration is below the limit specified by the magnetron manufacturer.

Finally, Figure 3.14 compares the simulated radiation patterns of the horn antenna with and without a metal lens in both E- and H-planes. Details within the beam 3-dB range are presented in Figure 3.15. As can be noticed, the lens enhanced the directivity in the main direction by 2.3 dB. The total gain of the radiating system is 18.1 dB.

It is important to mention that the presence of the lens affected the back radiation, which increased from −7 dB to 4.3 dB. However, the front-to-back radiation is 14 dB. The overall performance of the system was not affected by this fact.

Figure 3.16 shows the integration of the lens into the RIDAD setup. The planar lens was supported by a wooden frame that did not impact radiation characteristics.

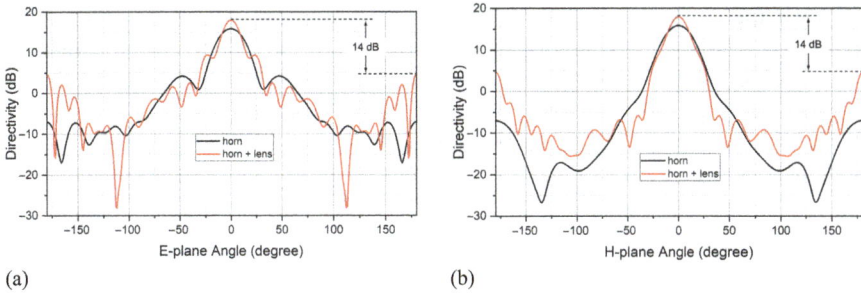

(a)   (b)

Figure 3.14   *Radiation pattern with (black) and without the metallic lens (red). (a) E-plane. (b) H-plane. The lens enhances the directivity in the main direction by 2.3 dB.*

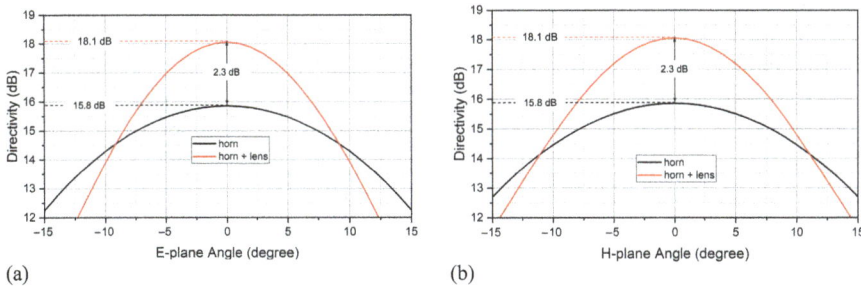

(a)   (b)

Figure 3.15   *Detail of the radiation pattern. With (black) and without the metallic lens (red). (a) E-plane. (b) H-plane. The lens enhances the directivity in the main direction by 2.3 dB.*

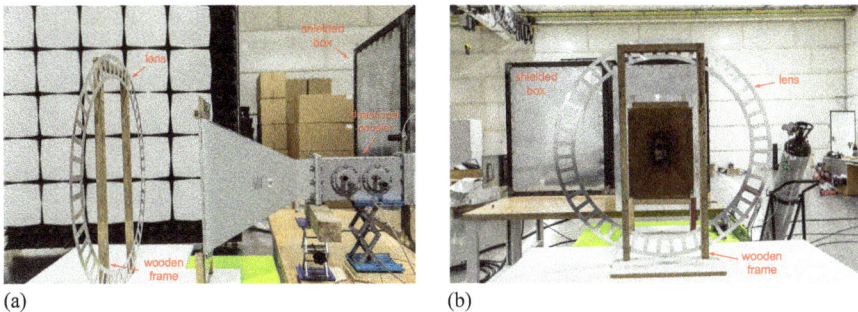

(a)   (b)

Figure 3.16   *Horn antenna and metallic lens integrated into the RIDAD setup. (a) Side view. (b) Front side.*

The system's final configuration, H- and E-planes, are elevation and azimuth directions.

3.8   Integration into a shielded box

Although the setup presented in Figure 3.2 allows a preliminary performance assessment, the system required the integration into a shielded box in order to prevent self-interference of the microwave signal with the numerous electronic sub-systems.

The shielded box included power filters, RF gaskets, refrigerant, air inlets, and waveguide port access.

The connection between the shielded box and the remote-control unit was achieved by a fiber optic link.

3.9   Measurements

The fields radiated by RIDAD were measured at the anechoic chamber of the Direct Energy Researcher Center, using a setup illustrated in Figure 3.17.

The measurement chain comprised a gain-calibrated LB780 horn antenna connected to a 6-GHz/4x40GS/s Teledyne oscilloscope, using an optical link.

The radiated electric field was measured at 6.8 m, 19 m, and 25 m of distance, as indicated in Figure 3.17. A picture of the measurement setup is presented in Figure 3.18. As it can be seen the RIDAD system is fully integrated within the shielding box.

Figure 3.19 presents a measured electric field, at 6.8 m. The measured peak amplitude is 8.1 kV/m and the duration of the pulse is 1.7 μs. The normalized field at 1 m is 55.1 kV/m.

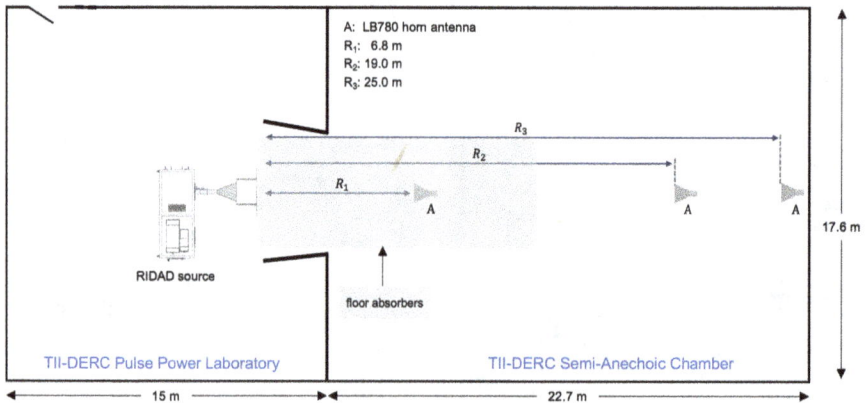

Figure 3.17   Radiation test setup

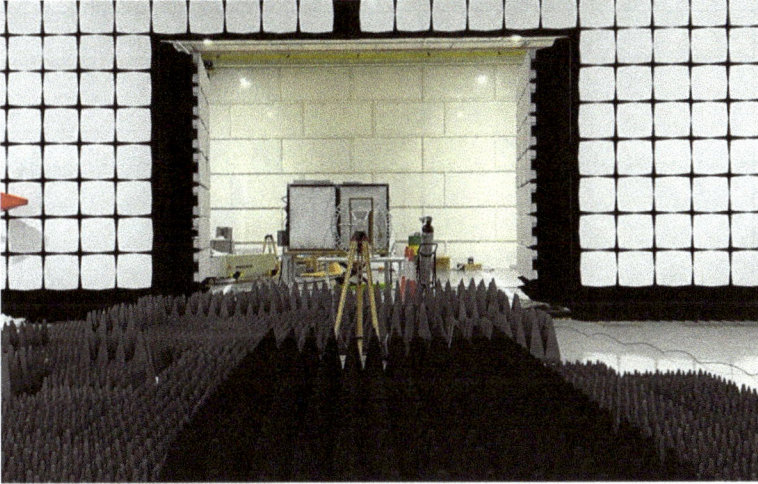

Figure 3.18   Front view of the RIDAD measurement set up in TII-DERC laboratories

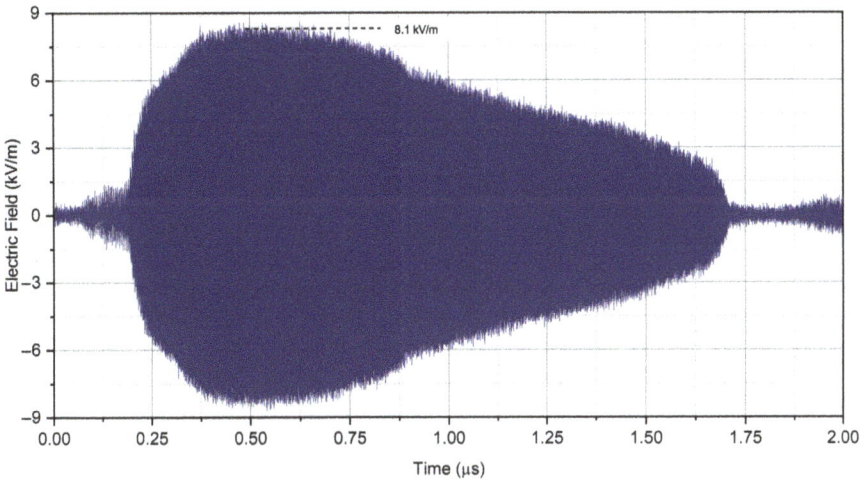

Figure 3.19   Measured time-domain electric field at 6.8 m from the RADIAD horn antenna

The measured field indicates that the peak power delivered by the magnetron to the radiating antenna is 1.5 MW.

Figure 3.20 shows the Fourier transform of the measured electric field. As it can be seen the fundamental frequency is 1.31 GHz. A minor secondary frequency component at 1.51 GHz also be seen.

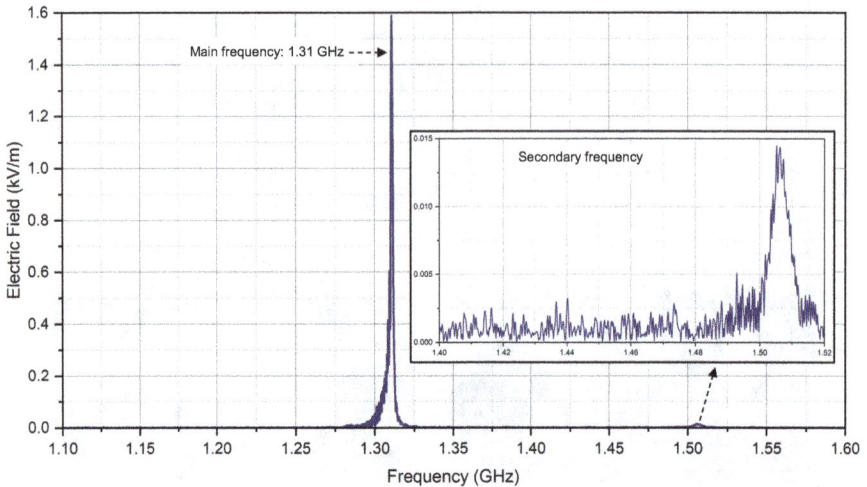

Figure 3.20   Fourier transform of the field measured at 6.8 m

The system was tested at pulse durations ranging from 1 μs to 3 μs, the measured peak fields were consistent with the results presented in Figures 3.19 and 3.20.

3.10   Conclusions

In this chapter, we have presented the integration of a High-Power Electromagnetic Radiator.

The integration work conducted involved the design and successful implementation of several subsystems, consisting of High Voltage power source, mode converter, cooling system, improved radiating system, and sensors and controls.

The system represents a cost-effective solution for EMC high-power electromagnetic testing, in lab conditions.

References

[1]   M. G. Bäckström, and K. G. Lövstrand, "Susceptibility of electronic systems to high-power microwaves: Summary of test experience," *IEEE Transactions on Electromagnetic Compatibility*, vol. 46, no. 3, pp. 396–403, 2004.

[2]   Tubedata, *Microwave Tube Operation* 1975 (altanatubes.com.br)

[3]   Raytheon RK65517/QK358, https://frank.pocnet.net/sheets/201/6/6517.pdf

[4]   Scandinova, https://scandinovasystems.com/pulse-modulator/m-series/m100-i-m100d-i/.

[5]   N. Wang, Q. Liu, C. Wu, L. Talbi, Q. Zeng, and J. Xu, "Wideband fabry-perot resonator antenna with two complementary fss layers," *IEEE Transactions on Antennas and Propagation*, vol. 62, no. 5, pp. 2463–2471, 2014.

[6]   D. Sanchez-Escuderos, H. C. Moy-Li, E. Antonino-Daviu, M. Cabedo-Fabres, and M. Ferrando-Bataller, "Microwave planar lens antenna designed with a three-layer frequency-selective surface," *IEEE Antennas and Wireless Propagation Letters*, vol. 16, pp. 904–907, 2017.

[7]   M. Cabedo-Fabres, E. Antonino-Daviu, A. Valero-Nogueira, and M. Ferrando-Bataller, "The theory of characteristic modes revisited: A contribution to the design of antennas for modern applications," *IEEE Antennas and Propagation Magazine*, vol. 49, no. 5, pp. 52–68, 2007.

[8]   Santillan-Haro, E. Antonino-Daviu, D. Sanchez-Escuderos, M. Ferrando-Bataller, "Design of high-gain antennas for 5G systems using characteristic modes," *IEEE International Symposium on Antennas and Propagation & USNC/URSI National Radio Science Meeting*, 8–13 July 2018, DOI:10.1109/APUSNCURSINRSM.2018.8608784.

[9]   Dassault Systems, CST Studio Suite, 2023.

[10]   D. M. Pozar (Ed.), *Microwave Engineering* New York: John Wiley & Sons, 2011.

[11]   C. A. Balanis (Ed.), *Advanced Engineering Electromagnetics* New York: John Wiley & Sons, 2012.

Chapter 4

Directed Energy Center at the University of New Mexico (DEC@UNM)

Edl Schamiloglu[1], Ganesh Balakrishnan[1,2], Brian Topper[1,2] and Alexander Neumann[2]

DEC@UNM (Directed Energy Center at the University of New Mexico) was established on October 21, 2021, through a cooperative agreement with the Air Force Research Laboratory (AFRL) titled Directed Energy Center for Lasers and Microwaves. This is a congressionally mandated center following a successful FY2021 UNM Federal Priority Request directed to the New Mexico delegation.

What makes DEC@UNM unique is that it is the only directed energy (DE) academic center in the USA performing state-of-the-art research in both DE lasers and DE microwaves.

4.1   DE—an introduction

The United States Department of Defense (DOD) defines DE weapons as those using concentrated electromagnetic (EM) energy, rather than kinetic energy, to "incapacitate, damage, disable, or destroy enemy equipment, facilities, and/or personnel." DE weapons include high-energy laser (HEL) and high-power microwave (HPM) weapons [1].

In general, HEL weapons potentially offer lower logistical requirements, lower costs per shot, and—assuming access to a sufficient power supply—deeper magazines compared with traditional munitions (Although a number of different types of HELs exist, many of the United States' current programs are solid-state lasers, which are fueled by electrical power. As a result, the cost per shot would be equivalent to the cost of the electrical power required to fire the shot.). These characteristics could, in turn, produce a favorable cost-exchange ratio for a defender, whose marginal costs would be significantly lower than those of an aggressor [1].

[1]Department of Electrical and Computer Engineering, University of New Mexico, Albuquerque, NM, United States
[2]Center for High Technology Materials, University of New Mexico, Albuquerque, NM, United States

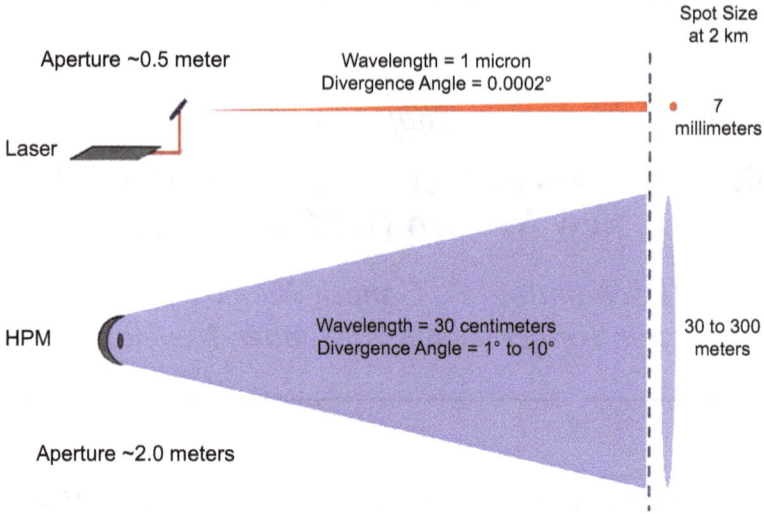

Figure 4.1   *A comparison of wave beam sizes for HELs and HPM devices (Source: Congressional Research Service image [1] based on an image in Mark Gunzinger and Chris Dougherty, Changing the Game: The Promise of Directed-Energy Weapons, Center for Strategic and Budgetary Assessments, April 19, 2021, p. 40, available from: https://csbaonline. org/uploads/documents/CSBA_ChangingTheGame_ereader.pdf. Note: Units of distance and beam widths are illustrative and are not to scale.)*

Similarly, HPM weapons provide a non-kinetic means of disabling adversary electronics and communications systems. These weapons could potentially generate effects over wider areas—disabling any electronics within their wave beam cone—than HEL weapons, which emit a narrower beam of energy (see Figure 4.1). Some analysts have noted that HPM weapons might provide more effective area defense against missile salvos and swarms of unmanned aircraft systems. HPM weapons in an anti-personnel configuration might provide a means of nonlethal crowd control, perimeter defense, or patrol or convoy protection [1].

Today, an increasing number of DE systems are being fielded so the technology is making its way out of the research laboratory into the hands of the warfighter. However, continued basic research is needed to improve current HEL and HPM systems, as well as enable more capabilities across the EM spectrum. This is consistent with the US DOD's Electromagnetic Spectrum Superiority Strategy [2].

DOD DE programs are coordinated by the Principal Director for Directed Energy within the Office of the Under Secretary of Defense for Research and Engineering (OUSD[R&E]).* The Principal Director for Directed Energy is

---

*Note, the description of the role of the Principal Director for Directed Energy and the DE roadmap are excerpted from [1].

responsible for the development and oversight of the HEL *Directed Energy Roadmap*, which articulates the DOD's objective of "[achieving] dominance in DE military applications in every mission and domain where they give advantage." According to OUSD(R&E), the current roadmap outlines DOD's plans to increase power levels of HEL weapons from around 150 kilowatt (kW), as is currently feasible, to around 300 kW by FY2023, "with goal milestones to achieve 500 kW class with reduced size and weight by FY2025 and to further reduce size and weight and increase power to MW [megawatt] levels by FY2026." For reference, although no consensus exists regarding the precise power level that would be needed to neutralize different target sets, DOD briefing documents (see Figure 4.2) suggest that a laser of approximately 100 kW could engage UASs, rockets, artillery, and mortars, whereas a laser of around 300 kW could additionally engage small boats and cruise missiles flying in certain profiles (i.e., flying across—rather than at—the laser). Lasers of 1 MW could potentially neutralize ballistic missiles and hypersonic weapons.

The authors of this chapter were unable to identify a similar roadmap for the HPM program. Perhaps this is because HPM technology is less mature and the budget is much smaller than the budget for the HEL program. Examples of recent HPM systems that have been fielded can be found in [1].

Most recently, in FY2023, the U.S. Office of the Secretary of Defense (OSD0 requested $16 million for High Energy Laser Research Initiatives, including basic research and educational grants, and $49 million for High Energy Laser Development, which funds applied research. OSD additionally requested $111 million in FY2023 for High Energy Laser Advanced Development, which is focused on "scaling the output power of DE systems to reach operationally effective

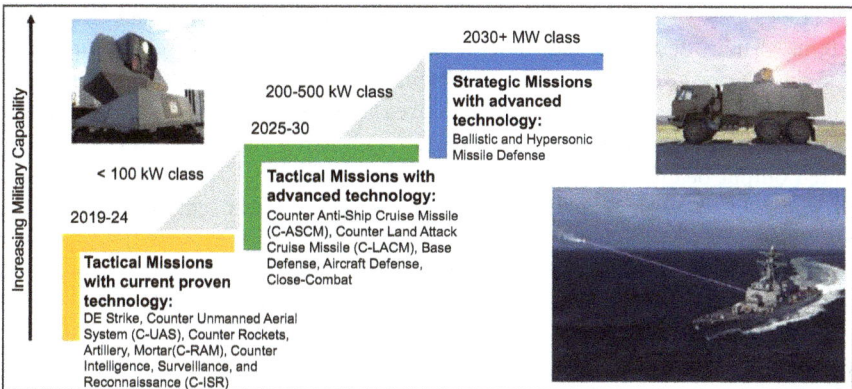

Figure 4.2   The HEL DE Roadmap [1] (Source: Dr. Jim Trebes, "Advancing High Energy Laser Weapon Capabilities: What is OUSD (R&E) (Office of the Undersecretary of Defense for Research and Engineering) Doing?," presentation at the Institute for Defense and Government Advancement (IDGA), October 21, 2020.)

power levels applicable to broad mission areas across the DOD." OSD requested $11 million in FY2023 to continue assessments of DE weapons, including assessments of the weapons' effects, effectiveness, and limitations. Finally, in the HPM arena, DARPA's Waveform Agile Radio-frequency Directed Energy (WARDEN) program seeks to "extend the range and lethality of high power microwave weapons ... [for] counter-unmanned aerial systems, vehicle and vessel disruption, electronic strike, and guided missile defense." DARPA received $20 million for WARDEN in FY2022 and requested $23 million for the program in FY2023.

The role of DEC@UNM is to: (1) advance the basic research to support DE lasers and microwaves; (2) train the next generation skilled technical workforce that would enter DOD and industry laboratories that are advancing the technology up the TRL [3] scale;[†] (3) disseminate its findings in peer-reviewed journals.

The next section describes the DE laser research that is ongoing at DEC@UNM. The following section describes the DE microwave research that is ongoing at DEC@UNM.

4.2   Laser research in DEC@UNM

The invention of the LASER (**L**ight **A**mplification by **S**timulated **E**mission of **R**adiation) in the early 1960s was followed by the research on MASERs (**M**icrowave **ASER**) the decade prior. Shortly thereafter, Elias Snitzer demonstrated the first fiber laser in 1964 [4]. Although between the 1960s and the 1990s tabletop laser systems consisting of bulk solid-state crystals dominated, by the 1990s, fiber lasers began to gain favor following maturation of the fiber technology in the telecommunications industry. By the end of the 20th century, output powers from fiber lasers reached the 100 W level [5].

Fiber laser systems have intrinsic benefits over bulk solid-state free space systems. Their compact and monolithic architecture allows robust operation, removing the necessity for cavity alignment and careful temperature and mechanical vibration control. The high surface area to volume ratio of the optical fiber makes heat removal efficient while the core-cladding design allows diffraction-limited output beams.

The 2000s saw an explosive growth in fiber laser power scaling. The kilowatt-level operation was achieved early [6] and 10 kW output at 1 micron was demonstrated before the end of the decade [7]. While beam combination has allowed multi-laser systems to provide ever-increasing output powers [8], the output power from a single fiber laser has mostly stalled since the first decade of the 21st century. This plateau has been due to newly discovered, at the time, issues of thermal

---

[†]Technology Readiness Levels (TRL) are a type of measurement system developed by NASA and are used to assess the maturity level of a particular technology. Each technology project is evaluated against the parameters for each technology level and is then assigned a TRL rating based on the projects progress [3]. There are nine technology readiness levels. TRL 1 is the lowest and TRL 9 is the highest. DEC@UNM performs basic research, which is at TRL 1. Upon maturation, some of the research may be considered TRL 2, but no higher.

nonlinearities [9]. The most prevalent thermal nonlinearity is known as thermal mode instability (TMI). The TMI describes the hopping between the desirable fundamental mode and higher-order modes in the core of the fiber at high temperatures. The result is unstable system operation and a reduction in the beam quality.

Significant research efforts have been made to understand and mitigate TMI. One of the more elegant solutions proposed was the use of anti-Stokes fluorescence (ASF) cooling [10]. ASF cooling is an all-optical technique of cooling a rare-earth doped gain medium by careful selection of the pump wavelength [11]. However, net cooling via ASF in ytterbium-doped silica—the material of choice for multi-kW level fiber laser systems—had never been observed. This changed in 2019 when researchers at UNM demonstrated ASF cooling of silica for the first time using an optical fiber preform [12].

In 2020, cooling by only several hundred milliKelvin was reported [13], making the experiments more of a neat academic trick. However, UNM researchers continued to push ASF cooling in silica in subsequent years [14,15]. The current world record for laser cooling ytterbium-doped silica by 67 Kelvin starting at room temperature was achieved in the DEC@UNM at the end of 2022 [16]. Experiments utilized a 100 W 1032 nm monolithic student-built fiber amplifier constructed in the same year. Now, the prospect of applying ASF cooling as a thermal mitigation technique in a monolithic fiber system seems increasingly achievable. Optimization of composition and fiber geometry is on the horizon at DEC@UNM, as commercially available fibers do not meet the profile suitable to implement ASF cooling designs in any practical way.

All contemporary kW-level fiber systems are in the 1 and 2-micron wavelength region [17,18]. A critical strategic wavelength range is the mid-infrared (MIR) atmospheric window spanning 2.5–5 microns. Fiber lasers in the tens of watts output level regime have been demonstrated in the 2.5–3 micron range [19,20]. The output powers of achieved longer wavelengths fall off exponentially. Even in the case of the former, output powers have yet to break 100 W. The current generation of these high-power fiber lasers operating around 3 microns are comprised of fluoride fibers and do not have the potential for single-fiber system power scaling beyond 100 W due to the low glass transition temperature of fluoride glasses.

As a result of these deep material limitations, new thermally stable materials need to be explored for future-generation MIR systems. To this end, researchers at DEC@UNM are currently working with Richard Weber's team at Materials Development, Inc. The collaboration has so far identified erbium lanthanum titanate glasses as strong candidates for MIR photonic applications. The lanthanum titanate glass has a transition temperature over 500 K above commonly employed fluoride materials while supporting mid-infrared fluorescence and transmission throughout the entire 2.5–5 micron atmospheric window. A future glass processing laboratory at DEC@UNM in the near future will continue to explore in-house other alternatives to fluoride and chalcogenide glasses that are more thermally, mechanically, and chemically durable.

4.3   High-power microwave research in DEC@UNM

The field of HPM emerged in the late 1960s and is a consequence of the emergence of pulsed power technology, initially in the United Kingdom, the United States, and the Soviet Union [21]. Pulsed power enabled the generation of relativistic, high-current electron beams which were then exploited for the generation of HPM radiation [22,23].

The University of New Mexico (UNM) established its program in DE microwaves in 1989. The program's initial focus was on the relativistic backward-wave oscillator (BWO) in X-band [24–29]. Then, in the early 2000s, the program shifted its focus to the relativistic magnetron in S- and L-band. In the 2010s research continued on the relativistic magnetron, expanding to the relativistic magnetron with diffraction output (MDO) [30–35]. In addition, the relativistic BWO was revisited, both in the context of a metamaterial source in the L-band and utilizing additive manufacturing to study a BWO in the E-band [36,37]. Furthermore, the use of a Bragg reflector instead of a cutoff neck or cavity reflector demonstrated conversion of the backward propagating TM_{01} mode to a forward propagating TE_{11} Gaussian wave beam [38].

Most recently, the thrust of the recent research activity has been on strategies to mitigate axial loss current in the relativistic magnetron and MDO [39–49]. This has been an ongoing collaboration with the group at Technion—Israel Institute of Technology, Haifa, Israel.

With the establishment of DEC@UNM, the thrust of the DE microwave research as part of this Center has been to advance the following: investigate a multi-stream traveling wave tube (TWT) amplifier in collaboration with the University of California-Irvine and expand the Center's research to focus in the X- to Ka-band frequency range.

4.3.1   Multi-stream TWT

HPM amplifiers are attractive since a broadband amplifier can lead to output signals with waveform diversity. Research on a multi-stream TWT is based on the linear theory developed by Prof. Alexander Figotin at the University of California at Irvine [50]. His theory, which is a fully Lagrangian field theory that reduces to the classic Pierce theory, suggests two important benefits of a multi-stream TWT:

1. Super-exponential growth rate (compared to exponential growth rate). As an example, his theory suggests that growth is $\exp az^n$ as opposed to $\exp az$ where n is close to 2.
2. Much wider bandwidth (note, this is the subject of current active investigation in order to quantify) than a TWT driven by a single beam.

In his book, Prof. Figotin states:

"The focal point of our efforts is the integration into the analytic model of the e-beam of multiple streams with different stationary velocities. The importance of such an integration is explained by the critical role played by electron velocities in the material medium-like behavior of plasma that differentiates it sharply from a fluid-like medium. D. Bohm and E. Gross wrote on this in their seminal paper [35, IV]:

... Thus, each particle suffers a wave-like perturbation in its velocity, which is larger for particles moving in the direction of the wave than for those moving in the opposite direction. The reason is that particles moving in the direction of the wave stay in phase longer and, therefore, experience a larger change of momentum in any given direction before the electric field, which imparts the momentum, is reversed. Particles going with the speed of the wave would stay in phase with the wave indefinitely, ...

In view of the above insight, it should come as no surprise that the multi-stream model of the e-beam is capable of accounting for more plasma features and with higher accuracy. The new features as compared to those in Pierce's theory ([50] – Sections [101, I], [105]) or its generalization include the following:

- space-charge (debunching) effects;
- multi-stream instability effects;
- complex instability frequency spectrum;
- rise of all-frequency instability modal branches with nearly constant phase velocities and consequently almost non-dispersive wave propagation;
- super-exponential amplification for pulses related to all-frequency instability modal branches.

As was mentioned earlier, we are actively studying the enhancement in bandwidth. The appearance of all-frequency instability modal branches, mathematically, infinite bandwidth. But this must be investigated further, which is what we are doing.

The first challenge in designing a multi-stream TWT is how to generate two beams with comparable currents and a 10%–20% energy difference from a single cathode stalk at a single potential. Early work in the late 1940s demonstrated a multi-stream TWT at low energies. In these early works, they were able to use two cathodes powered by two independent power supplies. This is not possible for pulsed power-driven sources of intense relativistic electron beams. The UNM group solved this matter and has moved on to design the X-band TWT amplifier [51,52].

4.3.2 The quest for higher frequency oscillators and amplifiers

There is currently tremendous interest in the community for HPM oscillators and amplifiers at higher frequencies than the community has previously been focused on. DEC@UNM is researching pulsed power-driven high-frequency HPM sources above the X-band. Because of the Pf^2 scaling of oscillators (where P is the output power and f is the frequency) and the $Pf\Delta f$ scaling of amplifiers (where Δf is bandwidth) [20,22], we expect that the power outputs from these sources will

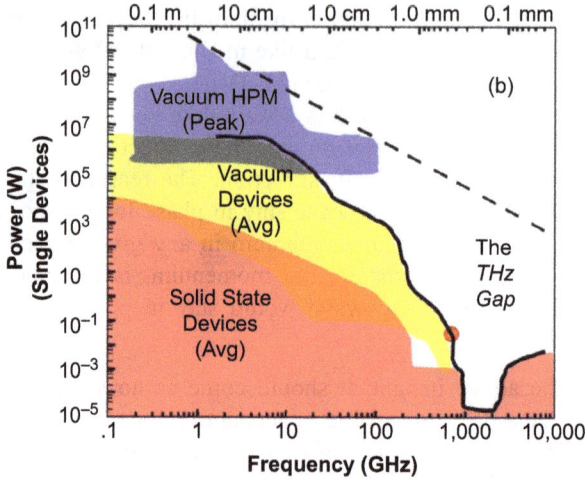

*Figure 4.3   Power vs. frequency of solid state and vacuum electronic devices. The
red circle data point represents a regenerative traveling wave tube
oscillator result [53].*

decrease as a function of frequency, as sketched out in the dashed line in Figure 4.3
(adapted from [54]).

We are presently at a nascent stage of this research and we will be reporting
our advances in the next few years. One example of a high frequency application is
the Active Denial System developed for crowd control and perimeter control [55].
It operates at 94 GHz.

4.4   Conclusions

DEC@UNM was established in October 2021 through a cooperative agreement with
the Air Force Research Laboratory (AFRL) titled *Directed Energy Center for Lasers
and Microwaves*. This is a congressionally mandated center following a successful
FY2021 UNM Federal Priority Request directed to the New Mexico delegation. It
was subsequently supported by an FY2023 UNM Federal Priority Request. What
makes DEC@UNM unique is that it is the only directed energy (DE) academic
center in the nation performing state-of-the-art research in both DE lasers and DE
microwaves. DEC@UNM is tasked with not only research and education towards
workforce development, DEC@UNM is leading the way by growing the directed
energy ecosystem in New Mexico. Learn more at dec.unm.edu.

Acknowledgments

Current DE laser research at UNM is being supported by the Air Force Office of
Scientific Research (FA9550-16-1-0362); and the Air Force Research Laboratory
Cooperative Agreement (FA9451-22-2-0016).

Current DE microwave research at UNM is being supported by AFRL Cooperative Agreement FA9451-22-2-0016, AFOSR MURI Grant FA9550-20-1-0409, and ONR Grant N00014-23-1-2072.

References

[1] U.S. Congressional Research Service, *Department of Defense Directed Energy Weapons: Background and Issues for Congress,* Report R46925, September 13, 2022, available from https://crsreports.congress.gov.

[2] U.S. Department of Defense, Electromagnetic Spectrum Superiority Strategy, October 2020.

[3] https://www.nasa.gov/directorates/heo/scan/engineering/technology/technology_readiness_level

[4] C.J. Koester and E. Snitzer, "Amplification in a fiber laser," *Appl. Opt.,* vol. 3, pp. 1182–1186, 1964.

[5] V.A. Dominic, S. MacCormack, R. Waarts, *et al.,* "110 W fibre laser," *Electron. Lett.,* vol. 35, pp. 1158–1160, 1999.

[6] Y. Jeong, J.K. Sahu, D.N. Payne, and J.A.N.J. Nilsson, "Ytterbium-doped large-core fiber laser with 1 kW continuous-wave output power," *OSA/ASSP 2004* (Optica Publishing Group,Washington, DC), 2004.

[7] C. Jauregui, T. Limpert, and A. Tünnermann, "High-power fibre lasers," *Nat. Photon.,* vol. 7, 861–867, 2013.

[8] A. Flores, T.M. Shay, C.A. Lu, *et al.,* "Coherent beam combining of fiber amplifiers in a kW regime," *CLEO 2011 (Science and Innovations, p. CFE3)* (Optica Publishing Group,Washington, DC), 2011.

[9] C. Jauregui, C. Stihler, and J. Limpert, "Transverse mode instability," *Adv. Opt. Photonics,* vol. 12, pp. 429–484, 2020.

[10] E. Mobini, M. Peysokhan, B. Abaie, M.P. Hehlen, and A. Mafi, "Spectroscopic investigation of Yb-doped silica glass for solid-state optical refrigeration," *Phys. Rev. Appl.,* vol. 11, p. 014066, 2019.

[11] M. Sheik-Bahae and R.I. Epstein, "Laser cooling of solids," *Laser Photonics Rev.,* vol. 3, pp. 67–84, 2019.

[12] E. Mobini, S. Rostami, M. Peysokhan, *et al.,* "Laser cooling of silica glass," arXiv preprint arXiv:1910.10609, 2019.

[13] E. Mobini, S. Rostami, M. Peysokhan, *et al.,* "Laser cooling of ytterbium-doped silica glass," *Commun. Phys.,* vol. 3, p. 134, 2020.

[14] B. Topper, M. Peysokhan, A.R. Albrecht, *et al.,* "Laser cooling of a Yb doped silica fiber by 18 Kelvin from room temperature," *Optics Lett.,* vol. 46, pp. 5707–5710, 2021.

[15] B. Topper, A. Neumann, A. Albrecht, *et al.,* "Potential of ytterbium doped silica glass for solid-state optical refrigeration to below 200 K," *Opt. Express,* vol. 3, pp. 3122–3133, 2023.

[16] B. Topper, A. Neumann, A. Albrecht, *et al.,* "Laser cooling silica: current status and future prospects," *Photonic Heat Engines: Science and Applications V, Proceedings SPIE,* vol. 10936, 12437, 6–10.

[17] B.M. Anderson, J. Solomon, and A. Flores, "1.1 kW, beam-combinable thulium doped all-fiber amplifier," *Fiber Lasers XVIII: Technology and Systems, Proceedings SPIE*, vol. 11665, 35–40.

[18] A. Flores, C. Robin, A. Lanari, and I. Dajani, "Pseudo-random binary sequence phase modulation for narrow linewidth, kilowatt, monolithic fiber amplifiers," *Opt. Express*, vol. 22, 17735–17744, 2014.

[19] V. Fortin, F. Jobin, M. Larose, M. Bernier, and R. Vallée, "10-W-level monolithic dysprosium-doped fiber laser at 3.24 µm," *Optics Lett.*, vol. 44, 491–494, 2019.

[20] V. Fortin, M. Bernier, S.T. Bah, and R. Vallée, "30 W fluoride glass all-fiber laser at 2.94 µm," *Optics Lett.*, vol. 40 2882–2885, 2015.

[21] E. Schamiloglu, R.J. Barker, M. Gundersen, and A.A. Neuber, "Modern pulsed power: Charlie Martin and beyond," *Proceedings of the IEEE*, vol. 92, 1014–1020, 2004.

[22] R.J. Barker and E. Schamiloglu, *High Power Microwave Sources and Technologies* (IEEE Press/John Wiley and Sons, New York, NY, 2001).

[23] J. Benford, J. Swegle, and E. Schamiloglu, *High-Power Microwaves*, 3rd Ed. (CRC Press, Boca Raton, FL, 2016).

[24] L.D. Moreland, E. Schamiloglu, R.W. Lemke, *et al.*, "Efficiency enhancement of high power vacuum BWOs using nonuniform slow wave structures," *IEEE Trans. Plasma Sci.*, vol. 22, 554–565, 1994.

[25] L.D. Moreland, E. Schamiloglu, R.W. Lemke, A.M. Roitman, S.D. Korovin, and V.V. Rostov, "enhanced frequency agility of high power relativistic backward wave oscillators," *IEEE Trans. Plasma Sci.*, vol. 24, 852–858, 1996.

[26] E. Schamiloglu, R. Jordan, M.D. Haworth, L.D. Moreland, I.V. Pegel, and A.M. Roitman, "High power microwave-induced TM01 plasma ring," *IEEE Trans. Plasma Sci.*, vol. 24, 6–7, 1996.

[27] F. Hegeler, C. Grabowski, and E. Schamiloglu, "Electron density measurements during microwave generation in a high power backward wave oscillator," *IEEE Trans. Plasma Sci.*, vol. 26, 275–281, 1998.

[28] C. Grabowski, E. Schamiloglu, C.T. Abdallah, and F. Hegeler, "Observation of the cross-excitation instability in a relativistic backward wave oscillator," *Phys. Plasmas*, vol. 5, 3490–3492, 1998.

[29] F. Hegeler, M. Partridge, E. Schamiloglu, and C.T. Abdallah, "Studies of relativistic backward wave oscillator operation in the cross-excitation regime," *IEEE Trans. Plasma Sci.*, vol. 28, 567–575, 2000.

[30] M. Fuks and E. Schamiloglu, "Rapid start of oscillations in a magnetron with a 'transparent' cathode," *Phys. Rev. Lett.*, vol. 95, p. 205101, 2005.

[31] M.I. Fuks and E. Schamiloglu, "70% efficient relativistic magnetron with axial extraction of radiation through a horn antenna," *IEEE Trans. Plasma Sci.*, vol. 38, 1302–1312, 2010.

[32] M. Liu, C. Michel, S. Prasad, M. Fuks, E. Schamiloglu, and C.-L. Liu, "RF mode switching in a relativistic magnetron with diffraction output," *Appl. Phys. Lett.*, vol. 97, p. 251501, 2010.

[33] C. Leach, S. Prasad, M. Fuks, and E. Schamiloglu, "Suppression of leakage current in a relativistic magnetron using a novel design cathode endcap," *IEEE Trans. Plasma Sci.*, vol. 40, 2089–2093, 2012.

[34] C. Leach, S. Prasad, M. Fuks, and E. Schamiloglu, "Compact relativistic magnetron with Gaussian radiation pattern," *IEEE Trans. Plasma Sci.*, vol. 40, 3116–3120, 2012.

[35] C. Leach, S. Prasad, M. Fuks, and E. Schamiloglu, "Compact A6 magnetron with permanent magnet," *Proc. 2012 IEEE International Vacuum Electronics Conference* (Monterey, CA, 24–26 April, 2012), pp. 491–492.

[36] S.C. Yurt, M.I. Fuks, S. Prasad, and E. Schamiloglu, "Design of a metamaterial slow wave structure for an o-type high power microwave generator," *Phys. Plasmas*, vol. 23, p. 123115, 2016.

[37] L. Bi, A. Elfrgani, and E. Schamiloglu, "E-band overmoded relativistic backward wave oscillator," Book of Abstracts IEEE PPPS-2019 (Orlando, FL, June 23–28, 2019).

[38] A. Elfrgani, S. Prasad, M.I. Fuks, and E. Schamiloglu, "Relativistic BWO with linearly polarized Gaussian radiation pattern," *IEEE Trans. Plasma Sci.*, vol. 42, 2135–2140, 2014.

[39] M.I. Fuks, S. Prasad, and E. Schamiloglu, "Efficient magnetron with a virtual cathode," *IEEE Trans. Plasma Sci.*, vol. 44, 1298–1302, 2016.

[40] M.I. Fuks, D.A. Andreev, A. Kuskov, and E. Schamiloglu, "Low energy state electron beam in a uniform channel," *Plasma*, vol. 2, 222–228, 2010.

[41] M.I. Fuks and E. Schamiloglu, "Application of a magnetic mirror to increase total efficiency in relativistic magnetrons," *Phys. Rev. Lett.*, vol. 122, p. 224801, 2019.

[42] D.A. Andreev, A. Kuskov, and E. Schamiloglu, "Review of the relativistic magnetron" *(Featured Article), Matter Radiat. Extremes*, vol. 4, p. 067201, 2019.

[43] J. Leopold, Y. Krasik, Y. Bliokh, and E. Schamiloglu, "Producing a magnetized low energy, high electron charge density state using a split cathode," *Phys. Plasmas*, vol. 27, p. 103102, 2020.

[44] J.G. Leopold, M.S. Tov, S. Pavlov, *et al.*, "Experimental and numerical study of a split cathode fed relativistic magnetron," *J. Appl. Phys.*, vol. 130, p. 034501, 2021.

[45] J.G. Leopold, Y. Hadas, Ya. E. Krasik, and E. Schamiloglu, "An axial output relativistic magnetron fed by a split cathode and magnetically insulated by a compact low power solenoid," *IEEE Trans. Electron Dev.*, vol. 68, 5227–5231, 2021.

[46] Ya.E. Krasik, J.G. Leopold, Y. Hadas, *et al.*, "An advanced relativistic magnetron operating with a split cathode and separated anode segments," *J. Appl. Phys.*, vol. 131, p. 023301, 2022.

[47] Y.P. Bliokh, Ya.E. Krasik, J.G. Leopold, and E. Schamiloglu, "Observation of the diocotron instability in a diode with a split cathode," *Phys. Plasmas*, vol. 29, 123901, 2022.

[48] J.G. Leopold, Y. Bliokh, Ya.E. Krasik, A. Kuskov, and E. Schamiloglu, "Diocotron and electromagnetic modes in split-cathode fed relativistic smooth bore and six-vane magnetrons," *Phys. Plasmas*, vol. 30, p. 013104, 2023.

[49] O. Belozerov, Y. Krasik, J. Leopold, *et al.*, "Characterizing the high-power-microwaves radiated by an axial output compact s-band a6 segmented magnetron fed by a split cathode and powered by a linear induction accelerator," *J. Appl. Phys.*, vol. 133, p. 133301, 2023.

[50] A. Figotin, *An Analytic Theory of Multi-Stream Electron Beams in Traveling Wave Tubes* (World Scientific, Hackensack, NJ, 2021).

[51] K.N. Islam and E. Schamiloglu, "Multiple electron beam generation with different energies and comparable currents from a single cathode potential for high power traveling wave tubes (TWTs)," *J. Appl. Phys.*, vol. 131, p. 044901, 2022.

[52] K.N. Islam, L.D. Ludeking, A.D. Andreev, S. Portillo, A.M. Elfrgani, and E. Schamiloglu, "Modeling and simulation of relativistic multiple electron beam generation with different energy from a single cathode potential for high power microwave sources," *IEEE Trans. Electron Dev.*, vol. 69, 1380–1388, 2022.

[53] J. Tucek, K. Kreischer, D. Gallagher, R. Vogel, and R. Mihailovich, in *Proc. 2007 IEEE International Vacuum Electronics Conf.* (Kitakyushu, Japan, 15–17 May 2007), pp. 219–220.

[54] J.H. Booske, "Plasma physics and related challenges of millimeter-wave-to-terahertz and high power microwave generation," *Phys. Plasmas*, vol. 15, p. 055502, 2008.

[55] See, for example, Joint Intermediate Force Capabilities Office, "Active Denial System FAQs," https://jnlwp.defense.gov/About/Frequently-Asked-Questions/Active-Denial-System-FAQs/.

Chapter 5

Compact and efficient mode converter for HPEM applications in L band

Luciano P. de Oliveira[1], Felix Vega[1], Abdul Baba[1], John J. Pantoja[2], Adamo Banelli[3] and Chaouki Kasmi[1]

5.1 Introduction

This chapter presents the mode converter's design, manufacturing, and integration for a pulsed Megawatt source in L-Band. This critical component in a high-power electromagnetics (HPEM) system [1] aims to maximize the power transfer between the microwave source and the transmission line connecting to the antenna, maximizing efficiency and minimizing standing waves while fulfilling size and weight requirements and optimizing the overall system performance [2–6].

In many cases, like the one presented here, the microwave source output and the antenna waveguide present different transversal sections, requiring impedance matching and mode matching to avoid refracting the energy, which can be catastrophic for the source.

The mode converter discussed here is designed to connect a custom-sized circular TE_{11}, ceramic-sealed waveguide into a standard WR650 TE_{10} waveguide insulated with air [7].

The mode converter's design is based on the relatively simple LC matching principle. However, in practical terms, the realization required several designs based on the principle of a rectangular waveguide and manufactured using CNC technology.

5.2 Mode converter design

Figure 5.1 shows the schematic of Remote Induction of Disturbance for Access Denial (RIDAD) 1, where a magnetron ET 6517 is connected to a horn antenna. As can be seen, the magnetron has a custom-size circular waveguide output, while the antenna is a standard WR650 horn antenna. The pulsed-magnetron output waveguide is insulated in ceramic with an unknown dielectric constant, while the horn

[1]Directed Energy Research Center, Technology Innovation Institute, Abu Dhabi, United Arab Emirates
[2]Single-Photon Group, School of Engineering and Physical Sciences, Heriot-Watt University, Edinburgh, UK
[3]NSI-MI Technologies, Tuscany, Italy

antenna input waveguide is insulated in air. Here, an additional challenge is presented. To achieve the desired return loss and mode matching levels, it is necessary to estimate the dielectric constant of the insulator, which is not specified in the magnetron datasheet. This estimation will aid in designing a mode converter that meets these requirements.

5.3   Dielectric constant estimation

Accurately estimating the dielectric constant is crucial in designing the mode converter, as it determines its load. The open-ended coaxial probe method was used to estimate the dielectric constant of the magnetron output waveguide insulator [8,9].

The open-ended coaxial probe method is used to measure the dielectric properties of materials. The method involves inserting a calibrated coaxial probe with a planar open termination with an extended ground plane, shown in Figure 5.2, into a sample material

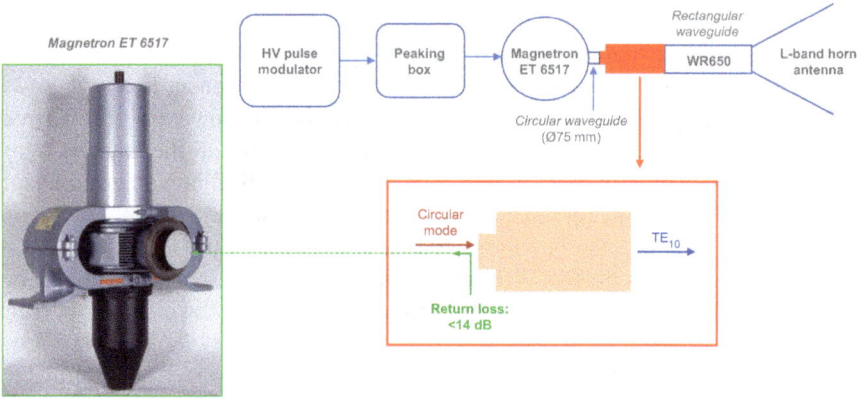

Figure 5.1   *Schematic of RIDAD—HPEM radiator prototype in L band*

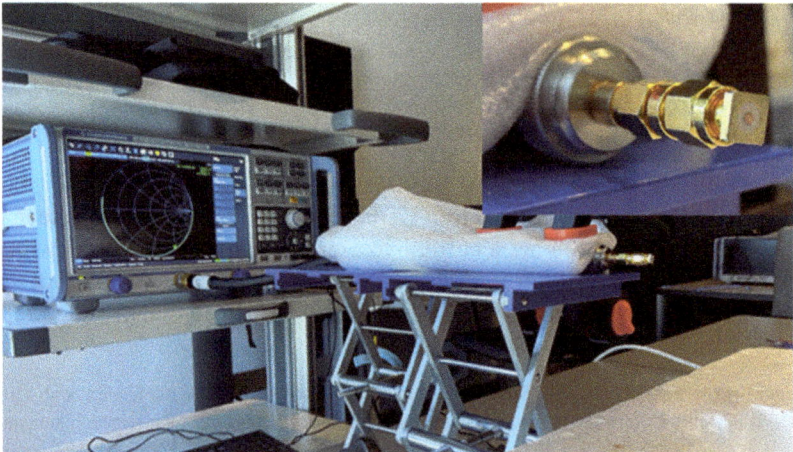

Figure 5.2   *Open-ended coaxial cable measurement setup*

under test (MUT) and measuring the reflection coefficient at the open end of the probe. The probe is connected to a calibrated vector network analyzer (VNA), and the reflection coefficient is measured by the VNA, which provides information about the dielectric properties of the material. The dielectric properties of the material can be calculated using a calibration process that involves measuring the reflection coefficient of a known standard material, such as air or solid material with well-characterized dielectric properties.

The magnetron output insulator measurements were done using a Rohde & Schwarz ZNB 40 VNA, as shown in Figure 5.2, over a 1.2 to 1.4 GHz frequency range. The data was post-processed according to [3], resulting in an average dielectric constant of about over the whole bandwidth.

5.4 TE11 to TE10 transition: mode mismatching

After measuring the magnetron output insulator dielectric constant, a numerical electromagnetic model was created to assess the impedance and mode mismatching at the WR650 and magnetron output. The numerical model is shown in Figure 5.3, where Port 1 and Port 2 indicate the magnetron and the WR650 feedings, respectively. The polarization at Port 1 was chosen according to the information provided by the magnetron datasheet.

Numerical simulation with CST Design Studio [10] showed that both waveguides propagate their respective fundamental modes over the target bandwidth, TE11 in the circular magnetron output waveguide and TE10 in the WR650. The modal E-field distributions, at 1.3 GHz, over plane AA' for the circular waveguide (TE11) and BB' for the WR650 (TE10) are presented in Figure 5.4. As can be seen, the mode mismatching can be solved by properly aligning the waveguides so the electric field TE10 and TE10 modes will be aligned.

The impedance matching, however, requires additional care, as shown in Figure 5.5. The measured impedance and the corresponding reflection coefficient at

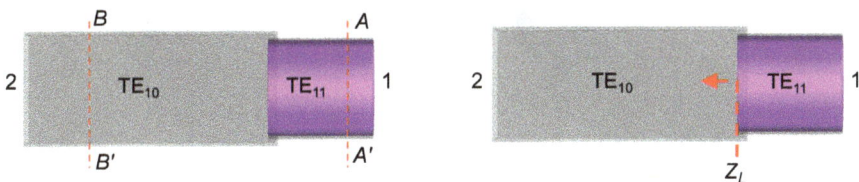

Figure 5.3   Magnetron direct transition to WR650

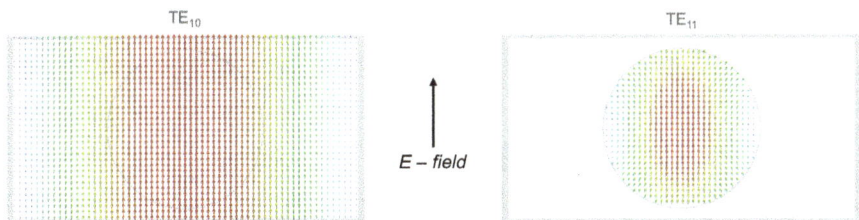

Figure 5.4   Modal E-field distribution inside the WR650 (left) and the circular waveguide (right) at 1.3 GHz

the transition were, respectively, $Z_L = 707.5 + 538.2$ Ω and $|\Gamma|_L = 0.71$, indicating that half the power is being reflected.

The next is to decide on the most suitable technique to match the circular and WR650 waveguide impedances. As shown in Figure 5.6, metallic irises and posts usually solve such a matching problem [11–13].

Both offer a compact matching transition and broadband impedance matching. Nonetheless, such solutions might present high-power limitations when the system operates at maximum power (internal breakdown).

A single waveguide stub, presented in Figure 5.7 [5], was chosen to address such a matching problem. The single stub is a robust solution for high-power

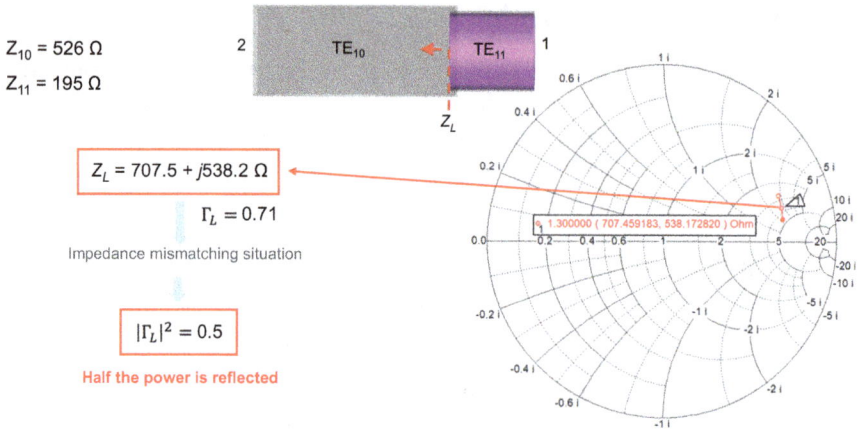

Figure 5.5   Direct transition mismatching

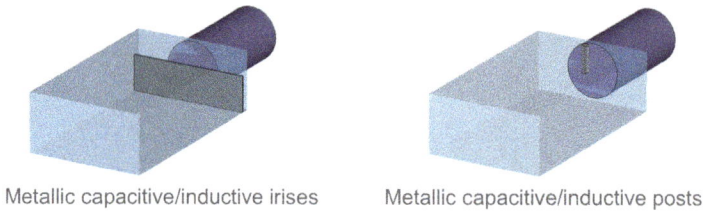

Metallic capacitive/inductive irises   Metallic capacitive/inductive posts

Figure 5.6   (Left) Metallic irises and (right) posts for waveguide matching impedance

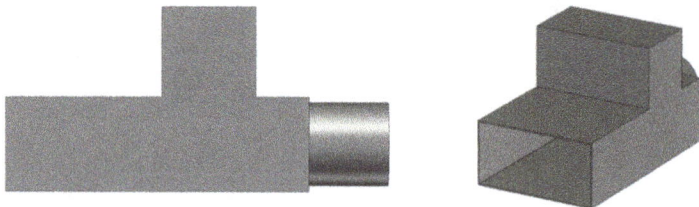

Figure 5.7   Single metallic waveguide stub

operation. Although offering a narrower bandwidth than the irises and posts, its effective bandwidth is enough for the system's operation.

5.5 Single metallic waveguide stub design

When the stub is correctly designed and positioned, it can act as a reflector of the microwave energy, causing a standing wave pattern in the waveguide. The amplitude and phase of the reflected wave can be controlled by adjusting the length and position of the stub, allowing for precise tuning of the system's impedance.

The single waveguide stub was designed using the classical approach, where the stub position and length are defined once the transmission line is loaded [11]. The stub position is defined by finding the position in the rectangular waveguide section, d_0, where the real part of the impedance is equal to the impedance of the propagating mode TE10, $Z_{10} = 526$ Ω, as illustrated in Figure 5.8.

By verifying the equivalent impedance along the rectangular waveguide, the first location where the impedance real part is about Z_{10}, is at $d_0 = 59.6$ mm from the transition. The equivalent impedance at this distance is $Z_d = 526.8 + j1,049$ Ω. Although other positions multiple of half-wavelengths apart from d_0 can be used to install the stub, d_0 was chosen to keep the mode converter as compact as possible.

Once the stub position is defined and, therefore, the reactance that needs to be compensated by the stub is defined, the next step is to calculate the stub length. Since a short-circuited transmission line presents a reactive input impedance, the short-circuited stub length must be calculated to offer an input impedance of about $Z_S = -j1,049$ Ω, as illustrated in Figure 5.9.

The short-circuited stub was modeled in CST Design Studio using a WR650 section. For a length of $h_0 = 105$ mm, an input impedance of about $Z_h = -j1,008$ Ω was found.

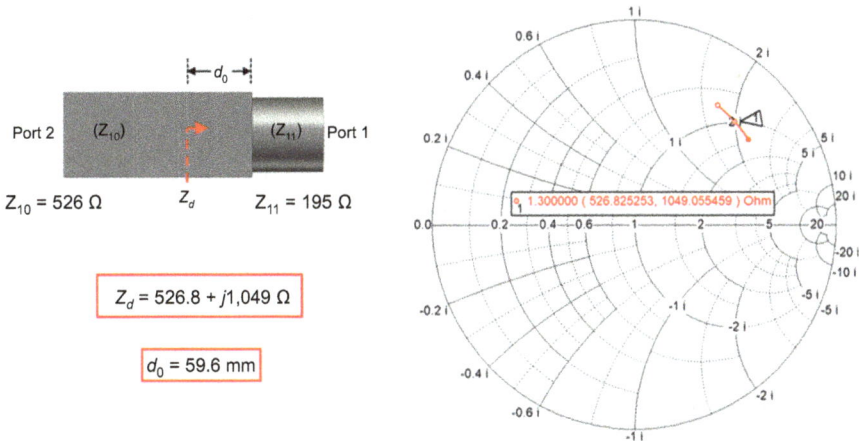

Figure 5.8 Stub insertion point position

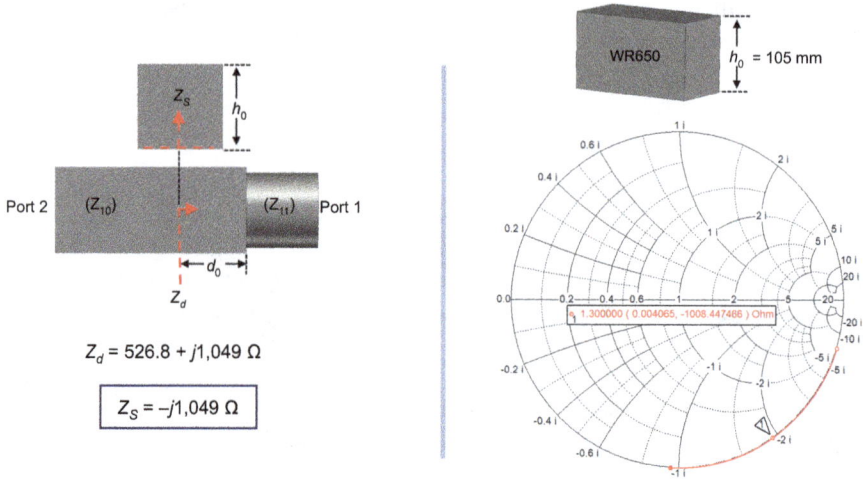

$Z_d = 526.8 + j1{,}049\ \Omega$

$Z_s = -j1{,}049\ \Omega$

Figure 5.9   Short-circuited stub length

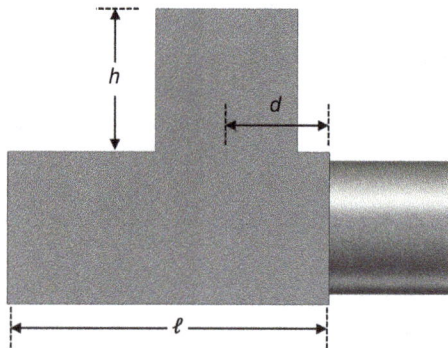

Figure 5.10   Mode converter model

Using these initial values—$d_0 = 59.6$ mm and $h_0 = 105$ mm, a numerical mode converter model consisting of a rectangular waveguide WR650 section (main transmission line), a WR650 short-circuited stub perpendicularly positioned to the main waveguide, and a circular waveguide filled with insulator was designed, as shown in Figure 5.10.

The final position and length of the stub—$d = 62.3$ mm and $h = 84.0$ mm, and the total length of the main waveguide, $l = 202.0$ mm, were adjusted to ensure the system's impedance tuning and the mode converter compactness, respectively. The return loss and insertion loss of the final mode converter design are presented in Figure 5.11. As can be seen, the mode conversion efficiency is higher than 97% over the frequency range from 1.29 to 1.31 GHz, with a return loss of less than

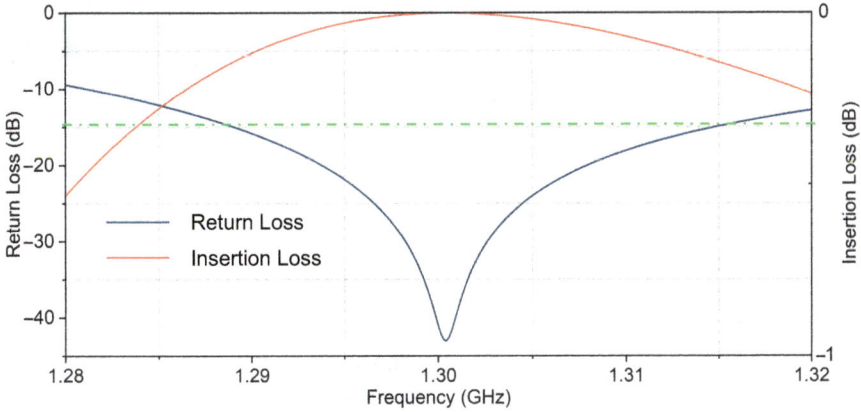

Figure 5.11   Mode converter's return loss and insertion loss

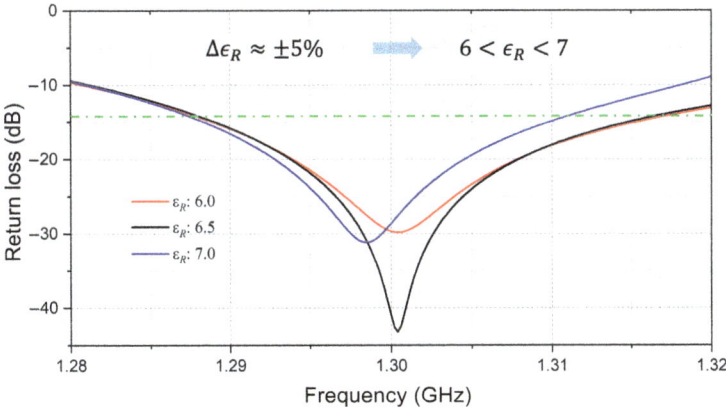

Figure 5.12   Effect of dielectric constant variation on the mode converter tuning

−14 dB over the bandwidth. The conversion efficiency is about 99.3% at the central frequency, with a return loss of approximately −42 dB at 1.30 GHz.

The final design was subjected to an analysis of variation in the insulator's dielectric constant—$6 < \varepsilon_R < 7$, to check how permittivity variation could affect the tune. The results, displayed in Figure 5.12, show that the mode converter response is quite robust around 1.3 GHz, with a return loss of around −14 dB over a frequency range from 1.29 GHz to 1.31 GHz.

5.6   Mode converter prototype

The final mode converter design comprises a rectangular waveguide WR650 (main transmission line) with one end connected directly to the circular waveguide through a custom-made flange on the transversal excitation plane. The remaining end was terminated on a standard rectangular flange, allowing the antenna to connect. Figure 5.13 shows the mode converter CAD model and the prototype. The primary waveguide was manufactured in aluminum on a CNC with a total length of 202 mm. The 84.7-mm long stub was placed on one of the broad walls at 62.3 mm from the rectangular end.

Due to a lack of access from the magnetron side, direct measurement of the mode converter's return and insertion losses was not feasible. Indirect measurements were carried out to confirm whether the prototype met the fabrication requirements. The measurement setup is presented in Figure 5.14. To verify the manufacturing accuracy, the de-embedded return loss phase was measured under two conditions: the mode converter output (Port 2) open and the mode converter output short-circuited. Then, the de-embedded measured return loss phases were compared to the corresponding simulated configuration.

Figure 5.13   Mode converter. CAD model (right). b) Prototype (left).

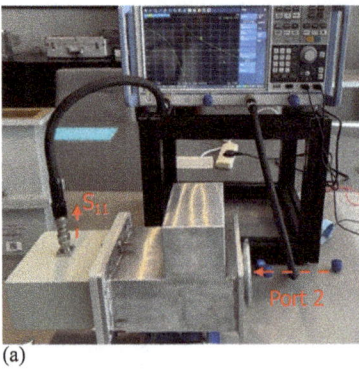

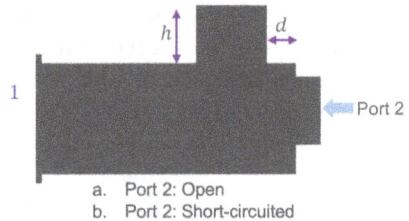

a.   Port 2: Open
b.   Port 2: Short-circuited

(a)   (b)

Figure 5.14   Mode converter measurement setup. (a) Prototype setup.
(b) Schematic setup.

Figure 5.15 compares the measured and simulated return loss phase when Port 2 is open and short-circuited. The simulated and measured results agree, indicating that the manufacturing tolerances met the precision requirements.

Due to the manufacturing technique, the narrow walls are not continuous and present slots, as seen in Figure 5.16. To evaluate the effect of these slots over the E-field, simulations were carried out in CST Design Studio, assuming an average input power of 1.3 MW and a maximum E-field of about 460 kV/m at the

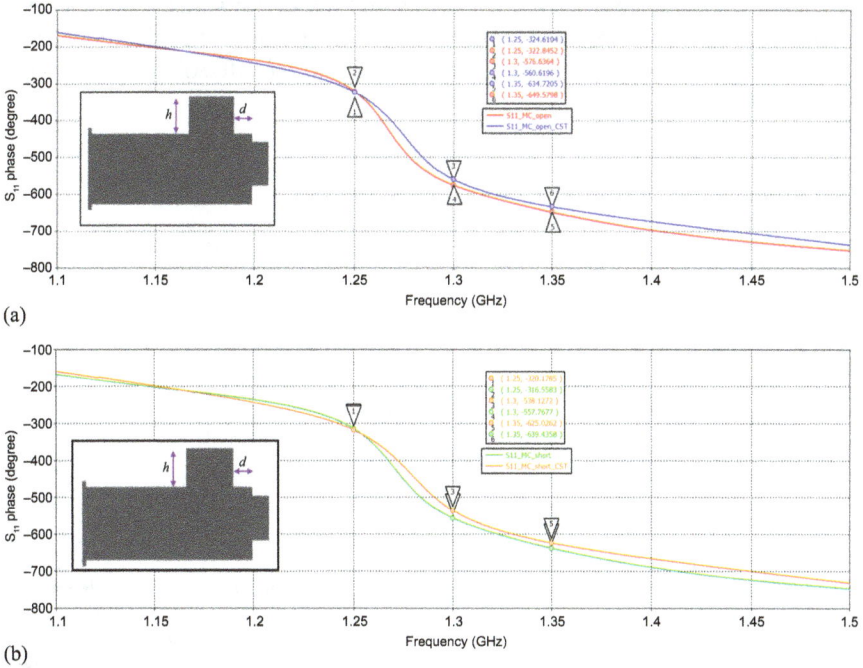

(a)

(b)

Figure 5.15 Simulated and measured return loss phase. (a) Open circular output. (b) Short-circuited circular output.

Figure 5.16 Mode converter prototype showing slots on the narrow walls

center of the main waveguide. As shown in Figure 5.17, the simulations did not
have a critical effect on electric performance. Aside from a small distortion on the
E-field distribution over the gap, the local maximum E-field does not exceed the
breakdown strength of air for all tested configurations. The return and insertion
losses also were not affected.

5.7   Integration into the magnetron and tests

After being validated through measurements and numerical simulations, the mode
converter prototype was installed in the RIDAD setup for the initial high-power tests.
Figure 5.18 shows the mode converter integrated into the RIDAD system, installed

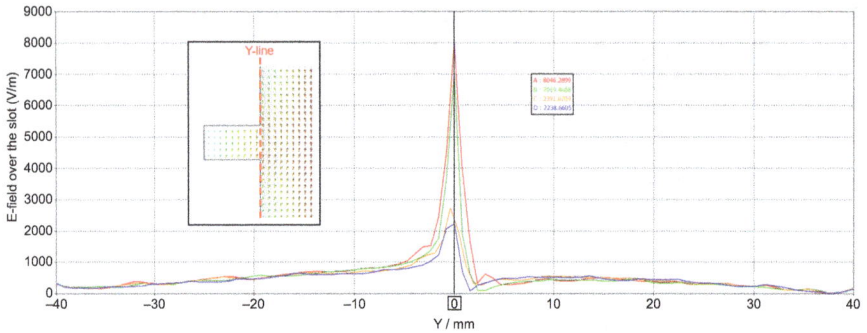

Figure 5.17   E-field on the narrow wall along the Y-line

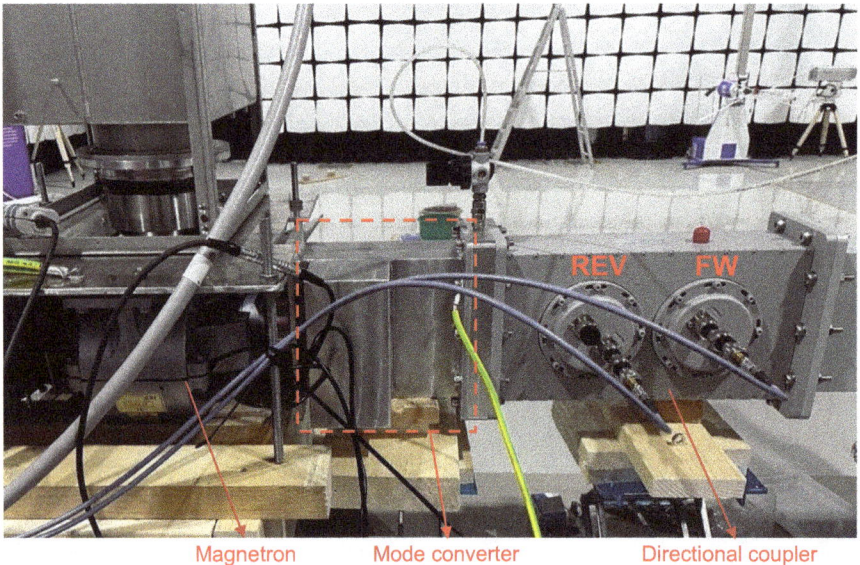

*Figure 5.18   Mode converter prototype integrated between the magnetron and the
directional coupler*

between the magnetron and the directional coupler. The directional coupler was installed during the tests to measure the power delivered by the magnetron. The power is sampled at the directional coupler forward port (FW) and transmitted to an 8-GHz/4x40 GS/s Teledyne oscilloscope.

Figure 5.19 presents the compensated time-domain voltage measured at the directional coupler forward port when the magnetron delivered about 1.3 MW. The voltage has a maximum value of 8 kV and a duration of about 1.5 µs. This voltage was then used to calculate the magnetron's power delivery.

Figure 5.20 presents the time-domain power estimated from the voltage in Figure 5.19, with a maximum of 1.28 MW, corresponding to about 98.4% of the power transmitted by the magnetron.

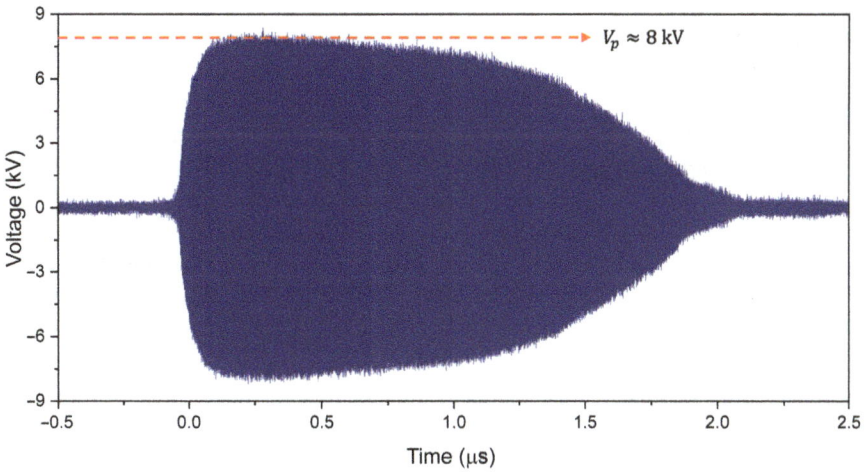

Figure 5.19   Time-domain voltage measured at the directional coupler forward port

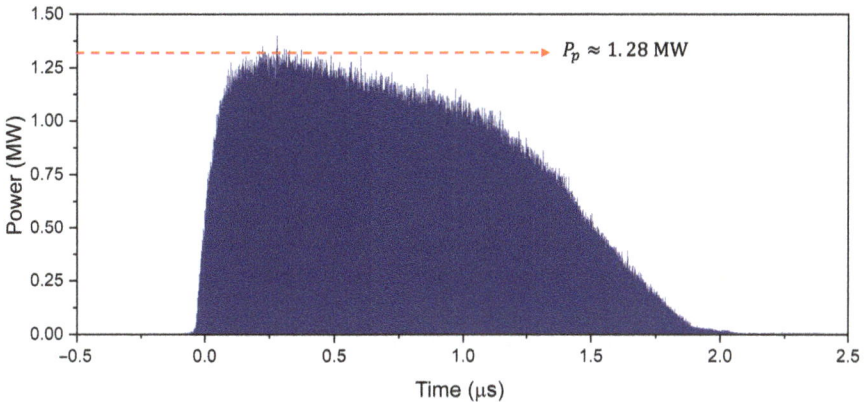

Figure 5.20   Time-domain power estimated from the measured voltage presented in Figure 5.19

5.8   Conclusion

This chapter describes the design of a mode converter that employs a custom-sized circular TE11 pulsed-magnetron output waveguide insulated with ceramic. The device is connected to a standard WR650 TE10 waveguide insulated with air. The mode converter was designed using the LC matching technique on a rectangular waveguide and manufactured with CNC technology.

The mode converter has a return loss of less than −20 dB within a 1.29 to 1.31 GHz bandwidth. This converter is integrated into the High-Power Electromagnetic Radiator (RIDAD). When the RIDAD operates at a nominal power of 1.3 MW, the mode converter outputs 1.28 MW, indicating a conversion efficiency of 98.4%.

References

[1]   Backstrom, M. G. and Lovstrand, K. G., "Susceptibility of electronic systems to high-power microwaves: Summary of test experience," *IEEE Trans. Electromagn. Compat.*, vol. 46, no. 3, pp. 396–403, 2004.

[2]   Lai, J. S. and Peng, F. Z., "Multilevel converters-a new breed of power converters," *IEEE Trans. Ind. Appl.* vol. 32, no. 3, pp. 509–517, 1996.

[3]   Barker, R. J. and Schamiloglu, E. (Eds.), *High-Power Microwave Sources and Technologies*, IEEE Press/Wiley, New York, NY, 2021.

[4]   Granatstein, V. L., and Alexeff, I., *High-Power Microwave Sources,* Artech House, London, 1987.

[5]   Liao, S. Y., *Microwave Devices and Circuits,* Prentice-Hall, Englewood Cliffs, NJ, 1996.

[6]   Benford, J., Swegle, J. A., and Schamiloglu, E., *High-Power Microwave,* Taylor & Francis, CRC Press, Boca Raton, FL, 2007.

[7]   de Oliveira, L. P., Vega, F., Baba, A. R., Pantoja, J. J., Banelli, A., and Kasmi, C., Compact and Efficient Mode Converter for HPEM Applications in L Band, *Presented at the GlobalEm 22*, Abu Dhabi, Abu Dhabi, UAE, 2022. [Online]. Available: https://www.globalem2022.com/assets/images/papers/62.pdf

[8]   Collin, R. E., *Foundations for Microwave Engineering*, 2nd Edition, John Wiley & Sons, Inc., Hoboken, NJ, 2001.

[9]   Riera, B. O., "Permittivity measurements using coaxial probes," *Barcelona Tech*, 2016.

[10] Dilman, I., Akinci, M. N., Yilmaz, T., Çayören, M., and Akduman, I., "A method to measure complex dielectric permittivity with open-ended coaxial probes," *IEEE Trans. Instrum. Meas.*, vol. 71, pp. 1–7, 2022, doi:10.1109/TIM.2022.3147878.

[11] CST Design Studio, https://www.3ds.com/products-services/simulia/products/cst-studio-suite.

[12] Pozar, D. M. (Ed.), *Microwave Engineering*, Wiley, New York, 2011.

[13] Gustrau, F., *RF and Microwave Engineering: Fundamentals of Wireless Communications*, John Wiley & Sons Ltd, Chichester, 2012.

Chapter 6

Design and characterization of the tapered impedance half impulse radiating antenna, TI-HIRA

Fernando Albarracin[1], Felix Vega[1], Adamo Banelli[2], Abdul Baba[1] and Chaouki Kasmi[1]

This chapter presents design considerations and experimental results of the tapered impedance half impulse radiating antenna (TI-HIRA) when fed by a commercial 50-Ω-pulsed generator. The TI-HIRA incorporates a tapered impedance transformer to the antenna feeding structure itself, simplifying the equipment sequence for the radiation of fast rise-time electromagnetic pulses. The characterization of a prototype and the measured electric field waveform are presented.

6.1   Introduction

The impulse radiating antenna (IRA) has been extensively utilized over the past decades for emitting high-amplitude electromagnetic pulses commonly used in intentional electromagnetic interference (IEMI) and electromagnetic compatibility (EMC) testing and research [1]. The half-IRA (HIRA) is the monopolar version of the IRA and is implemented by introducing a ground plane in the horizontal symmetry plane [2–4]. In the conventional concept of operation of a HIRA, a constant-impedance TEM transmission line composed of coplanar-plate feeders illuminates a parabolic metal dish with a spherical wave, to ultimately collimate an intense pulse on boresight. The characteristic impedance of the feeding TEM line is constrained by the compromise between optimal illumination of the dish and feeder geometry, limiting the HIRA to impedance values close to 100 Ω [5]. For applications using a commercial unbalanced 50 Ω pulser as a generator, an external wideband impedance transformer to feed the HIRA is required. To simplify the required device sequence in the pulse radiation process, the authors introduced the tapered impedance HIRA (TI-HIRA) [2] as an alternative. The concept of the TI-HIRA consists of integrating a TEM feeder whose characteristic impedance

[1]Directed Energy Research Center, Technology Innovation Institute, Abu Dhabi, United Arab Emirates
[2]NSI-MI Technologies, Tuscany, Italy

changes as a function of the radial distance from the focal point to the dish aperture. Different impedance progression functions will result in slightly different TEM feeding arm geometries [4]. Since the aperture field distribution is kept unchanged, the waveform of the field radiated by the TI-HIRA is expected to be the same as that obtained when an external impedance adapter is connected between a conventional HIRA and the generator [2,4].

Related work reporting the development of conformal feeding structures for HIRA is available in the literature. A conformal feed geometry inspired by the Vivaldi opening was proposed and analyzed in [6] for a reflector IRA. Simulated results based on generators with output impedance higher than 100 Ω were presented with limited radiation efficiency, due to the phase center spatial dispersion. Another conformal feeding structure, based on an asymptotic conical dipole (ACD) to implement a UWB feeding structure was introduced in [7]. Various charge distributions for the feeder are derived numerically to implement the desirable input impedance. Baum *et al.* [8] utilized a set of shaped conical arms as the feeder for an 85 Ω HIRA. Such design is part of the well-known JOLT system, one of the most representative high-voltage transient radiators developed so far.

This chapter focuses on the design and experimental verification of the TI-HIRA, introduced in [2], in terms of impedance response and radiated field waveform. The integration of the novel tapered impedance feeder approach allows a direct connection of a 12 kV peak, 50 Ω commercial off-the-shelf (COTS) generator to the TI-HIRA. The chapter is organized as follows: the tapered impedance feeder concept for the TI-HIRA design is presented in Section 6.2. The implementation and experimental characterization of a TI-HIRA prototype is presented in Section 6.3. Section 6.4 provides conclusions and future lines of work.

6.2   The TI-HIRA concept

6.2.1   Classic HIRA design

In its conventional geometry, as the one depicted in Figure 6.1, the expected electric field waveform radiated from the HIRA can be computed using the expression first derived by Mikheev in [9], as follows:

$$e_{rad}(t,R) = f_1 \frac{1}{2\pi f_g} \cdot \left(\frac{v_{in}(t - R/c)}{R} \frac{\sin(\beta_0)}{1 + \cos(\beta_0)} - \cdots \right.$$

$$\frac{v_{in}(t - l/c - R_2/c)}{R_2} \frac{\sin(\beta_0) + \sin(\gamma)}{1 + \cos(\beta_0 - \gamma)} - \cdots$$

$$\left. \frac{4}{D} v_{in}(t - 2F/c - R/c) + (2 + 2\cos(\gamma)) \frac{v_{in}(t - l/c - R_2/c)}{D} \right) \left[\frac{V}{m} \right] \quad (6.1)$$

where f_1 is $0.25\sqrt{2}$, D and F are the diameter and focal length of the reflector, respectively. c is the speed of light in free space. The parameters l, β, γ, and R_2, are depicted in 1.1. This expression is valid only on boresight, and for near and far-field regions.

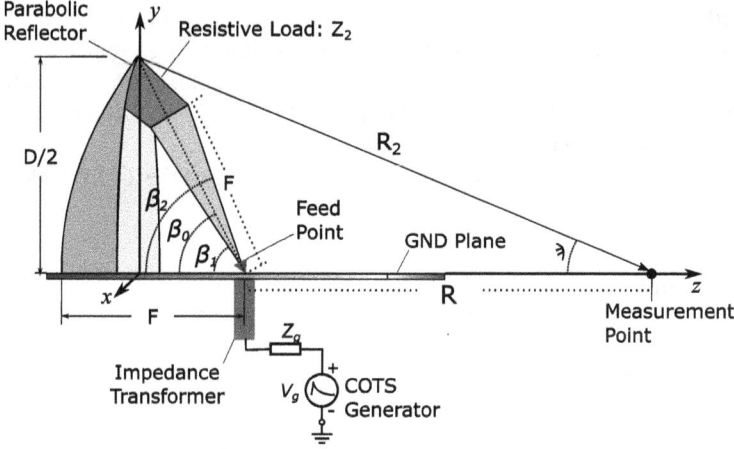

Figure 6.1   Conventional HIRA setup including a UWB, high-voltage impedance transformer between the 50 Ω generator and the feed point of the antenna

The geometry of the feeding arms is linked to the impedance of the TEM line illuminating the reflector with a spherical wave. The characteristic impedance of a single, constant-impedance, balanced, two-arm coplanar feeder [10] can be computed as:

$$Z_{TEM} = 120\pi \frac{K(m)}{K(1-m)} = 120\pi f_g \qquad (6.2)$$

where $K(m)$ is the complete elliptic integral of the first kind [11], and fg is the impedance geometrical factor. The parameter m is solved from (2) and will define the constitutive angles of the feeding arms.

The feeder impedance is related to the radiated field strength. The power normalized gain G_p has been defined by Farr in [10] as a radiation performance metric. It is defined as the ratio of the effective length of the IRA, h_a, to the square root of the impedance geometric factor f_g, as

$$G_p = \frac{h_a}{\sqrt{f_g}} = 2\pi c \sqrt{f_g} \frac{\|re_{rad}(t,r)\|}{\left\| \frac{dv_{in}(t)}{dt} \right\|} \qquad (6.3)$$

where h_a is computed as

$$h_a = \frac{D}{2} \frac{\pi m^{1/4}}{2K(1-m)} \left(1 - \frac{2}{\pi} \sin^{-1}\left(\frac{1-\sqrt{m}}{1+\sqrt{m}}\right)\right) \qquad (6.4)$$

The maximum value of the power gain, $G_p = 0.91$ (Figure 5.6 in [10]) corresponds to $m = 0.266$, which derives into a TEM feeder geometry with a characteristic impedance $Z_{TEM} = 301$ Ω.

In most practical implementations [12], Z_{TEM} is selected as 400 Ω for the full IRA case, where two pairs of coplanar constant-impedance TEM feeders are connected in parallel: $Z_{IRA} = Z_{TEM}||Z_{TEM} = 200\ \Omega$. This configuration eases the feeding of the full IRA by means of a wideband balun with an impedance ratio of 4:1 (200 Ω to 50 Ω), using two sections of 100 Ω–RF cable, allowing the IRA to be connected to a 50 Ω generator. For this practical case ($Z_{TEM} = 400\ \Omega$) G_p slightly drops to 0.89, corresponding to $m = 0.565$.

Once the value of m is found, the constitutive angles of the TEM feeders can be computed as:

$$\beta_0 = \tan^{-1}\left(\left(2(F/D) - \frac{1}{8(F/D)}\right)^{-1}\right), \tag{6.5}$$

$$\beta_1 = 2\tan^{-1}\left(m^{0.25}\tan\left(\frac{\beta_0}{2}\right)\right), \tag{6.6}$$

$$\beta_2 = 2\tan^{-1}\left(m^{-0.5}\tan\left(\frac{\beta_1}{2}\right)\right). \tag{6.7}$$

The analysis presented above applies also to the HIRA design, with the exception of the single-feeder impedance expression, Z_{TEM}, becomes now

$$Z_{TEM-HIRA} = \frac{1}{2}Z_{TEM-IRA} = 60\pi\frac{K(m)}{K(1-m)} = 200\Omega. \tag{6.8}$$

Following the quasi-optimal TEM feeding impedance approach, the commonly implemented input impedance for the HIRA is obtained by connecting two single feeding arms in parallel, thus, $Z_{HIRA} = Z_{TEM-HIRA}||Z_{TEM-HIRA} = 100\ \Omega$. Due to the presence of the ground plane, which halves the TEM impedance of the feeder (see Figure 6.1 as a reference), Z_{HIRA} is single-ended, usually implemented with a coaxial termination.

A system view of the conventional HIRA when connected to a single-ended generator through an impedance transformer is depicted in Figure 6.2(a). The radiated electric field can now be computed, in the frequency domain, as the cascaded transformer-HIRA system response:

$$E_{Rad}(f, R) = T_T(f)\frac{1}{l_{eff}(f, R)}V_g(f) \tag{6.9}$$

where T_T is the voltage transfer function of the impedance transformer and $1/l_{eff}(f, R)$ is the inverse of the effective length of the HIRA, computed as the Fourier transform of (1); which reads

$$\frac{1}{l_{eff}(f, R)} = f_1\frac{1}{2\pi f_g}\cdot\ldots\left(\frac{1}{R}\frac{\sin(\beta)}{1+\cos(\beta)}e^{-if\frac{R}{c}} - \frac{1}{R_2}\frac{\sin(\beta) + \sin(\gamma)}{1+\cos(\beta-\gamma)}e^{-if\frac{(l+R_2)}{c}}\right.$$

$$\left.-\frac{4}{D}e^{-if\frac{(2F+R)}{c}} + (2+2\cos(\gamma))e^{-if\frac{(l+R_2)}{c}}\frac{1}{D}\right)(m^{-1}) \tag{6.10}$$

(a)

(b)

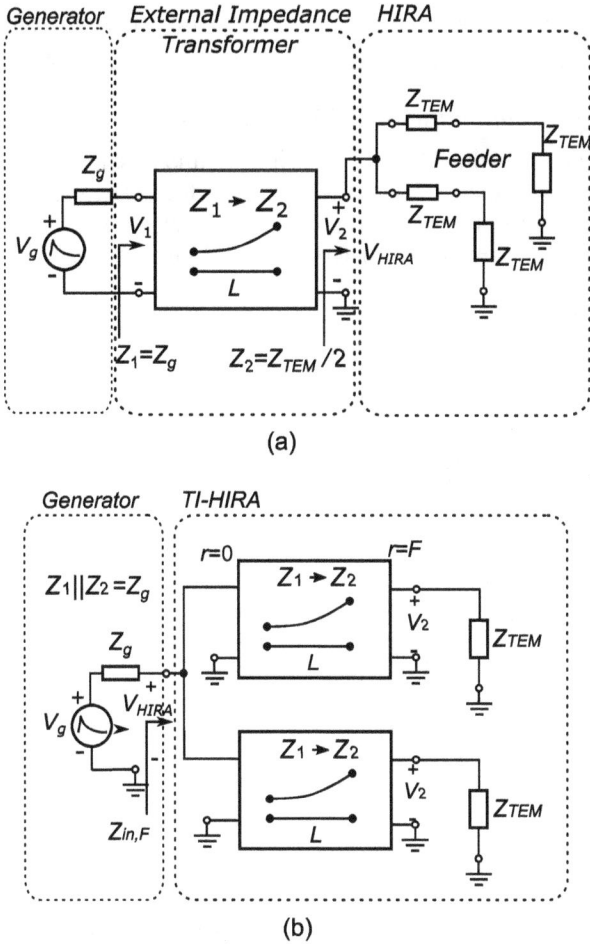

Figure 6.2 Circuit model of (a) the classical HIRA and (b) the TI-HIRA, when connected to a single-ended 50 Ω generator [2]

By means of the ABCD parameters [13], also known as the "chain" parameters, T_T can be computed as:

$$T_T(f) = \frac{Z_2}{Z_1} \cdot \frac{1}{C(f)Z_2 + D + A\frac{Z_2}{Z_1} + \frac{B}{Z_1}},$$ (6.11)

where $Z_1 = Z_g$ and $Z_2 = Z_{TEM-HIRA}/2$ are selected in practical implementations (see Figure 6.2(a)).

Multiple tapered impedance transformers have been proposed and extensively analyzed and characterized in the literature [13–15]. Various impedance progression functions can be implemented, and their ADCB parameters are either found in closed form or can be numerically computed to obtain T_T. An external impedance transformer with an exponential progression function, used to feed a HIRA was presented in [16].

6.2.2   Tapered impedance HIRA (TI-HIRA)

The design concept in the TI-HIRA consists of implementing the tapered impedance transformers in the feeding arms geometry itself. The system scheme of the TI-HIRA is depicted in Figure 6.2(b). The radiated electric field can also be computed as in (9), however, for the TI-HIRA case, T_T is computed as:

$$T_F(f) = \frac{V_2(f)}{V_g(f)} = \frac{Z_2}{A_F(f)Z_2 + B_F(f)} \cdot \frac{Z_{in,F}(f)}{Z_g + Z_{in,F}(f)} \cdot e^{(j2\pi fF/c)}. \qquad (6.12)$$

In this configuration, two impedance tapers are now connected in parallel. The term $\exp(j2\pi fF/c)$ accounts for the phase difference related to the traveling time of the signal through the impedance transformer in the classical HIRA case, implicitly included in (6.11). $Z_{in,F}$ is the impedance of the TI-HIRA seen at the apex of the feeders, computed as

$$Z_{in,F}(f) = \frac{1}{2}\frac{A(f)Z_{TEM} + B(f)}{C(f)Z_{TEM} + D(f)} \qquad (6.13)$$

The design starts with the selection of the impedance progression, as a function of the radial distance from the focal point of the reflector to the end of the feeding arms (see Figure 6.3 as a reference). The characteristic impedance of one of the

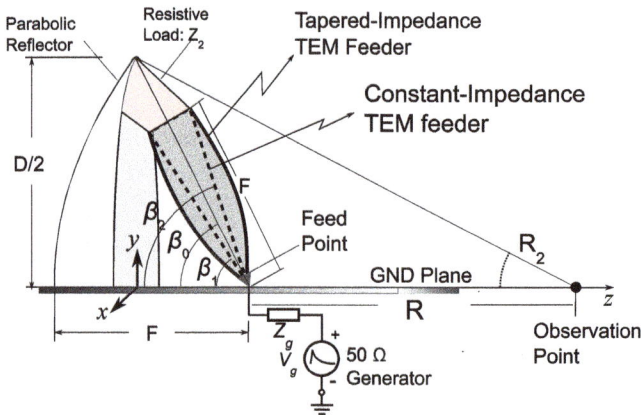

Figure 6.3   TI-HIRA geometry, showing a direct connection to a 50-Ω-pulsed generator. Angles β_0, β_1 and β_2 progressively change as a function of the distance from the focal point to the reflector, leading to the conformal shape of the TI-HIRA arms. The same angles have fixed values in the classical HIRA (slotted arms).

feeding arms is then computed as

$$Z_F(r) = f_Z(r) = 60\pi \frac{K(m(r))}{K(1 - m(r))}; \ Z_1; \text{ for } r = 0 Z_2; \text{ for } r = F \qquad (6.14)$$

where r is the radial distance, $Z_F(r)$ is the characteristic impedance of a single feeder, $f_Z(r)$ is the impedance progression function. Z_1 and Z_2 are, respectively, the initial and ending TEM impedance values. Note that the geometric factor m is now a function of the radial distance r too.

Figure 6.3 shows the TI-HIRA geometry directly connected to a 50 Ω generator. The conformal shape of the TI-HIRA feeding arms illustrates the progression of the feeder impedance from the focal point ($r = 0$) towards the reflector. The difference $\beta_2(r) - \beta_1(r)$ in degrees (see Figure 6.3) converges to the same value in the classical HIRA at $r = F$.

Numerous impedance progression functions, including exponential [14], logarithmic, Gaussian [13], and Klopfenstein [17] are feasible options. A detailed analysis of those impedance functions has been presented by the authors in [2]. Figure 6.4 shows the impedance progression function as a function of the r. The Klopfenstein function presents "unmatched" impedance values at $r = 0$ and $r = F$, however, lower reflections along the frequency are expected from it. Figure 6.5 shows the arm geometry for each of the analyzed tapering impedance functions. As a proof of concept, the exponential impedance function is presented and implemented for laboratory tests, as it will be shown in the remainder of this work.

The expression to compute $Z_F(r)$ for the exponential function case is as follows

$$Z_F(r) = Z_1 e^{\alpha r}; \alpha = \frac{1}{F} \ln \left(\frac{Z_2}{Z_1} \right) \qquad (6.15)$$

where α is the exponential factor, $Z_1 = 100 \ \Omega$ and $Z_2 = Z_{TEM} = 200 \ \Omega$.

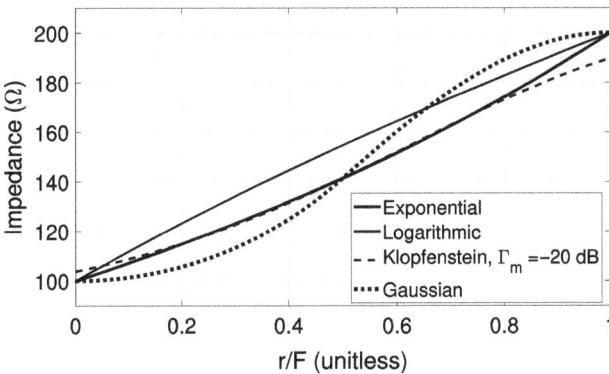

Figure 6.4   Impedance variation for the four feeder tapering impedance function vs. normalized feeder length

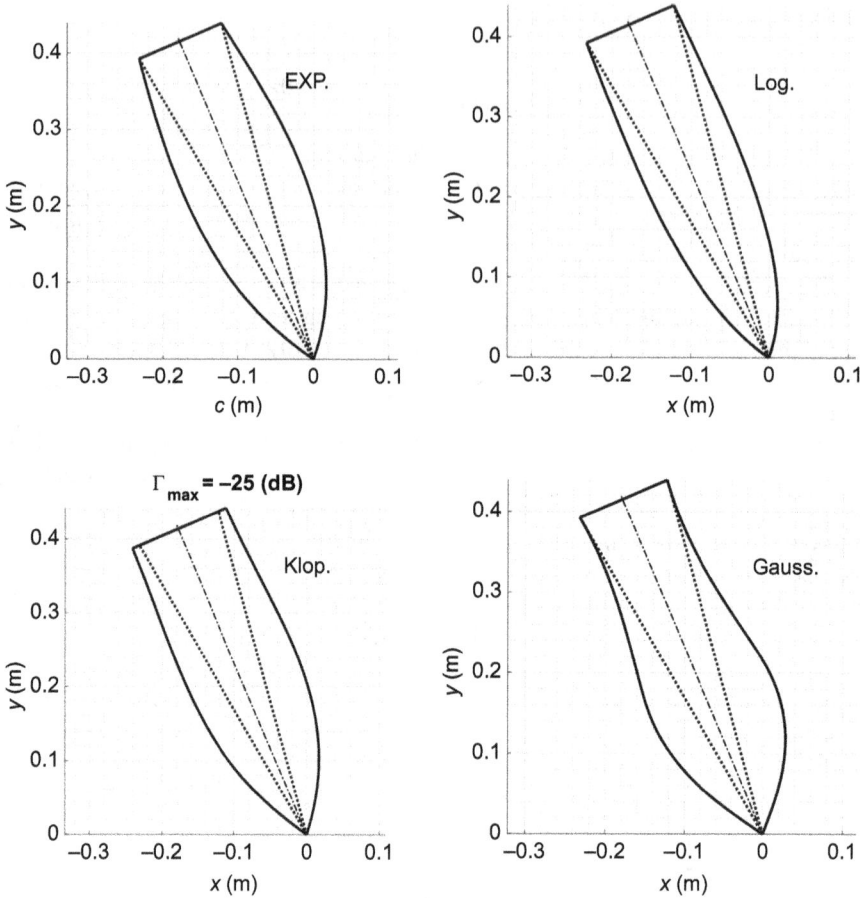

Figure 6.5   Alternative feeding arms geometries for the exponential, logarithmic, Klopfenstein, and Gaussian impedance functions

6.3   Practical implementation

A parabolic reflector dish, 1.2 m diameter and with 0.456 m in focal length was selected to implement the first prototype of the TI-HIRA, as shown in Figure 6.6. The values of Z_1 and Z_2 were selected as 100 Ω and 200 Ω, respectively, with the exponential progression shown in Figure 6.4. The conformal feeding arms follow a shape like the one shown in the top left of Figure 6.5. An arrangement of series and parallel carbon-based resistors, enclosed with acrylic for mechanical support were fabricated to terminate each feeding arm. The aim of this prototype is to feed the TI-HIRA directly from a 12 kV, 50 Ω, COTS generator. The rise-time of the voltage output is $t_r = 100$ ps with a full-width at half maximum (FWHM) of 3 ns [18].

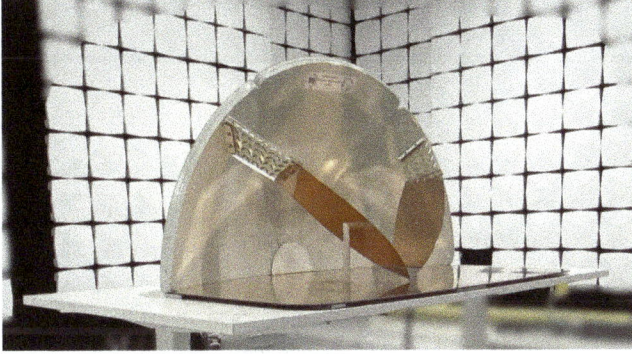

Figure 6.6   Prototype of the TI-HIRA with an exponentially tapered feeder

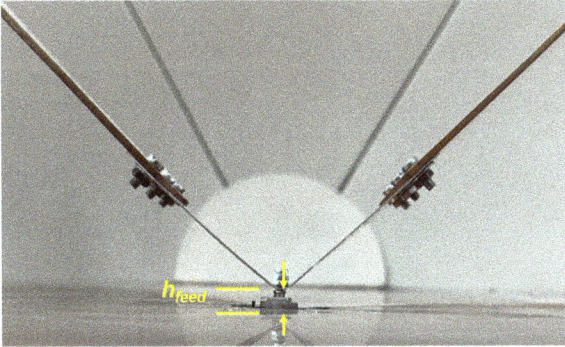

Figure 6.7   Detailed view of the feed transition at the focal point of the TI-HIRA

Special attention was paid to the transition between the HN connector and the feeding arms apex. A compromise between impedance response and voltage hold-off capability must be addressed in this part of the design. Figure 6.7 shows a detailed view of the feeding structure. The separation between the apex and the ground plane was gradually modified to reduce the discontinuity in the impedance transition at the TI-HIRA feeding point, thereby minimizing pulse distortion. The time domain reflecto-metry (TDR) response of the TI-HIRA impedance was measured for different values of the separation parameter (i.e., h_{feed} in Figure 6.7), and the results shown in Figure 6.8.

The instrument bandwidth was limited to 4 GHz for this measurement. A significant capacitive coupling between the connector and the feeding apex is identified from the slits in the TDR for $h_{feef} = 4.6$ mm and 5.6 mm for the case where the 2 mm thick arm-plates were directly connected to each other. The smoothest response is observed at $h_{feef} = 8$ mm. An aluminum "butterfly" shaped structure was finally integrated to the feeding point to minimize the area parallel to the ground plane, as shown in Figure 6.7.

Finally, the electric field waveform of the TI-HIRA was measured in a semi-anechoic chamber modified with absorbing material in the surrounding ground area

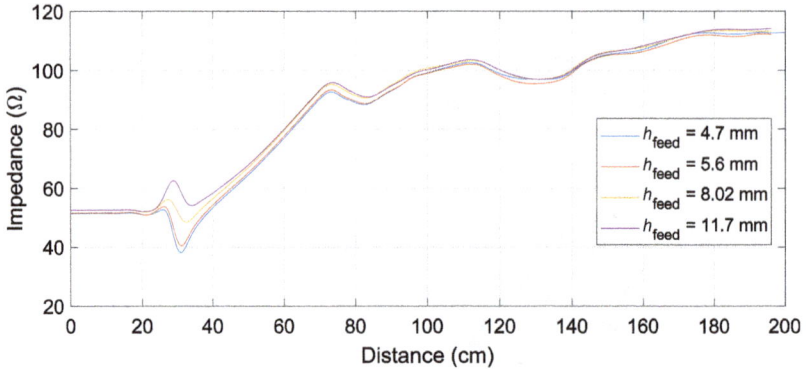

Figure 6.8   TDR response of the TI-HIRA with different gap values at the feeding point (see Figure 6.7 as a reference)

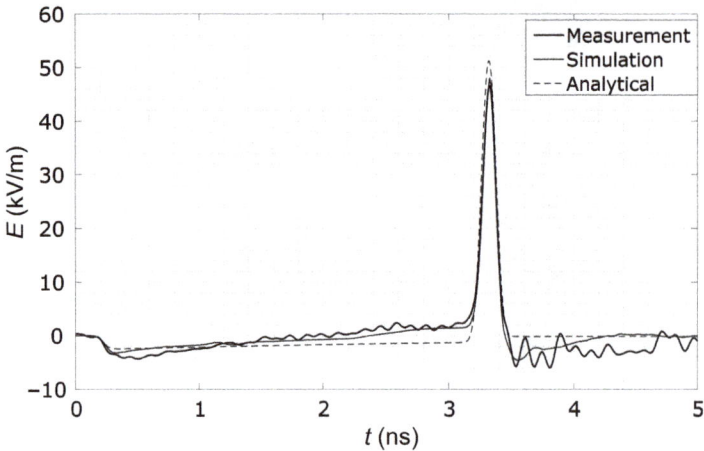

Figure 6.9   Radiated electric field waveform from the TI-HIRA, reduced to 1 m

of the antenna. On the receiving side, a free-field magnetic flux sensor, B-dot, whose equivalent area is $9 \times 10^{-6} \text{m}^2$, and 10 GHz cutoff frequency was used. A 10 GHz ultra-wideband balun connected between the B-dot and a 6 GHz oscilloscope completes the measurement setup. A shielded enclosure was used to protect the oscilloscope during the radiation tests. The shielding level ranges from 60 dB to 20 dB for frequencies between 100 MHz and 3 GHz. A UPS was used to avoid coupling through the main grid cable.

A comparison of the measured and the simulated electric field waveform on boresight, reduced to 1 m is presented in Figure 6.9. The purpose of such computation is to find the far-voltage figure of merit of the TI-HIRA. The effects of cabling and connectors have been compensated in the post-processing of the captured data. Small discrepancies can be observed in the prepulse area. However, the agreement between the calculated, measured, and simulated values is excellent.

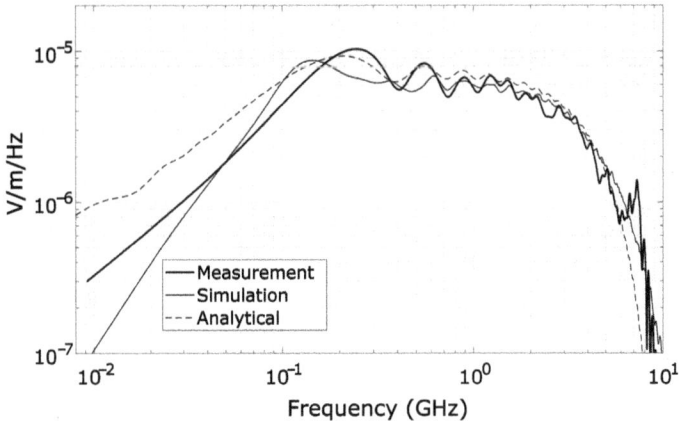

Figure 6.10 Measured spectral response of the electric field from the TI-HIRA

The measured peak electric field on boresight reached 47 kV/m, slightly lower than the simulated one (48 kV/m). The rise time and the pulse width of the measured waveform are approximately 120 ps and 200 ps respectively. The prepulse magnitude reached −4 kV/m and −3.25 kV/m for the measured and the simulated cases, respectively.

The spectral response is shown in Figure 6.10. Using the analysis presented in Section 6.2, an estimation of the radiated field, computed as the inverse Fourier transform of (6.9), gives a peak value of 51 kV/m, reduced to 1 m. A similar value is found by direct calculation of Eq. 6.1, using a double exponential voltage waveform with the same characteristics as those of the COTS generator used in the test. It is worth mentioning that this analytical value does not consider the beam squint from the edge of the finite ground plane.

6.4   Conclusion

The analysis, design, and practical implementation of the recently introduced tapered-impedance IRA, TI-HIRA, has been presented. The conventional HIRA design process has been complemented with a new approach in which the impedance transformer is part of the radiating structure. The presented prototype of the exponentially-tapered TI-HIRA was verified to perform in excellent agreement with the analytical and the simulated values.

The developed TI-HIRA allows direct connection of single-ended 50 Ω COTS pulsed generators, which are progressively more available in the market. The suppression of an external impedance transformer, apart from saving space, reducing costs, and manufacturing time, contributes to simplifying the transmitting system in applications involving radiation of fast-rising high-voltage pulses.

Open lines of research in this matter have been identified and should encourage further investigation and development. Practical implementation of other impedance functions is one example. Further exploration of the response of the tapered impedance profiles when excited by low-frequency sources deserves attention in future work as well.

References

[1]   Baum CE. Radiation of Impulse-Like Transient Fields. *Sensor and Simulation Note 321*; 1989. pp. 1–28.

[2]   Vega F, Albarracin-Vargas F, Kasmi C, *et al.* The Tapered Impedance Half-Impulse Radiating Antenna. *IEEE Transactions on Antennas and Propagation*. 2021;69(2):715–722.

[3]   Vega F, Albarracin F, Kasmi C, *et al.* Electric Field Radiation from the Tapered-Impedance Half Impulse Radiating Antenna. In: *GlobalEM*; 2022.

[4]   Albarracin F, Vega F, Baba AR, *et al.* Influence of the Impedance Profile on the Performance of the Tapered-Impedance Half-Impulse Radiating Antenna. In: *2023 IEEE International Symposium on Antennas and Propagation and USNC-URSI Radio Science Meeting (USNC-URSI)*; 2023. pp. 1731–1732.

[5]   Farr EG, and Baum CE. Prepulse Associated with the TEM Feed of an Impulse Radiating Antenna. Sensor and Simulation Note 337; 1992. pp. 1–36.

[6]   Manteghi M, and Rahmat-Samii Y. A Novel Vivaldi Fed Reflector Impulse Radiating Antenna (IRA). In: *2005 IEEE Antennas and Propagation Society International Symposium*. vol. 1A; 2005. pp. 549–552.

[7]   Singh DK, Pande DC, and Bhattacharya A. Improved Feed Design for Enhance Performance of Reflector Based Impulse Radiating Antennas. Sensor and Simulation Note 565; 2013. pp. 1–15.

[8]   Baum CE, Baker WL, Prather WD, *et al.* JOLT: A Highly Directive, Very Intensive, Impulse-Like Radiator. *Proceedings of the IEEE*. 2004;92(7): 1096–1109.

[9]   Mikheev OV, Podosenov SA, Sakharov KY, *et al.* New Method for Calculating Pulse Radiation From an Antenna With a Reflector. *IEEE Transactions on Electromagnetic Compatibility*. 1997;39(1):48–54.

[10]   Farr E. Optimizing the Feed Impedance of Impulse Radiating Antennas Part I: Reflector IRAs. Sensor and Simulation Note 354; 1993. pp. 1–37.

[11]   Abramowitz M, and Stegun IA. *Handbook of Mathematical Functions: With Formulas, Graphs, and Mathematical Tables*. North Chelmsford, MA: Courier Corporation; 2012. Google-Books-ID: KiPCAgAAQBAJ.

[12]   Giri DV. *Swiss Half IRA (SWIRA) Design Considerations*. Spiez, Switzerland; 2005.

[13]   Pozar DM. *Microwave Engineering*. New York: John Wiley & Sons; 2011.

[14]   Collin RE. The Optimum Tapered Transmission Line Matching Section. *Proceedings of the IRE*. 1956;44(4):539–548.

[15]   Dworsky LN. *Modern Transmission Line Theory and Applications*. Malabar, FL: Krieger Publishing Company; 1988.

[16]   Vega F, Rachidi F, Mora N, *et al.* Design, Realization, and Experimental Test of a Coaxial Exponential Transmission Line Adaptor for a Half-Impulse Radiating Antenna. *IEEE Transactions on Plasma Science*. 2013;41(1):173–181.

[17]   Klopfenstein RW. A Transmission Line Taper of Improved Design. *Proceedings of the IRE*. 1956;44(1):31–35.

[18]   KI Ltd. PBG3 Pulsed Sources. Available from: http://www.kentech.co.uk/index.html?/PBG3.html2.

Chapter 7

Cathode edge effect and divergence of emitted electron beams in vircators

Moza Mohamed[1] and Ernesto Neira[1]

Virtual Cathode Oscillators (Vircators) are High-Power Microwave Sources (HPMS) that have drawn significant interest for their potential applications in high-power microwave systems. Despite their promising potential, Vircators are considered inefficient in terms of energy performance, with the divergence of emitted electrons being one of the reasons for this. To address this matter, this chapter studies electron beam divergence from Vircators and how the Cathode Edge Effect impacts it. The objective is to explore the underlying mechanisms and provide insights for improving the Vircator's performance and energy efficiency.

This chapter is organized into four sections. Section 7.1 presents the operating principle of Vircators and the theoretical aspects behind beam divergence. Section 7.2 introduces the numerical techniques available for beam divergence simulations and presents one example of a study on beam divergence of cathodes with different rim sizes. In Section 7.3, experimental techniques measuring the beam divergence and the beam homogeneity are presented. Finally, the main findings are summarized in the conclusions in Section 7.4.

7.1 Vircators operating principles

Virtual cathode oscillator (vircator) is a class of HPMS [1] capable of providing gigawatt order peak power output at typical frequencies from 1 GHz to 10 GHz [2]. Owing to its simple mechanical structure and ability to operate without a magnetic guide [3], this enhances the device's desirability in many applications. Nevertheless, these sources exhibit low efficiency (typically less than 10% of peak output power) and sensitivity to gap closure that limits the operation time [4]. Currently, there are several configurations of Vircators [1,2], where the most utilized are the Axially Extracted and Reflex Triode (see Figure 7.1).

The operating principle of the Vircator can be explained by dividing it into two areas: the diode region and the drift region [5].

[1]Directed Energy Research Center, Technology Innovation Institute, Abu Dhabi, United Arab Emirates

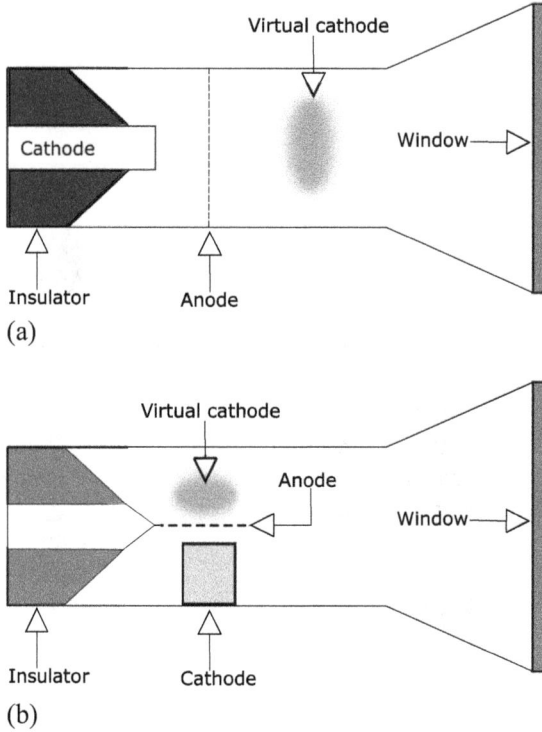

Figure 7.1   Most commonly utilized Vircator configurations. (a) Axially extracted vircator and (b) reflex triode.

7.1.1   Diode region

The diode comprises the cathode, the anode, and the region in between. Here, electrons emitted by the cathode are accelerated towards the anode due to the applied Voltage (see Figure 7.2). The electron emission mechanism taking place is known as Explosive Electron Emission (EEE) and can be summarized as follows: When a high voltage is applied to the electrodes, this causes a high electric field on the cathode surface. The cathode's micro-protrusions accumulate a large negative charge density. This results in a local electric field enhancement that can be up to 10^4. Due to Joule heating, the micro-protrusions quickly increase in temperature, leading to thermionic electron emission. As a consequence of the high electron density, an explosion occurs on top, creating an electron-ionized neutral cloud and generating plasma. This leads to intense electron emission, with current densities reaching 10^6–10^8 A/cm^2. This phenomenon, known as EEE, leads to plasma's rapid development and expansion over the cathode surface at 1–2 cm/μs typically, providing excess electrons for space charge limited flow [6,7]. Figure 7.2 shows a schematic of the diode region and the EEE phenomenon.

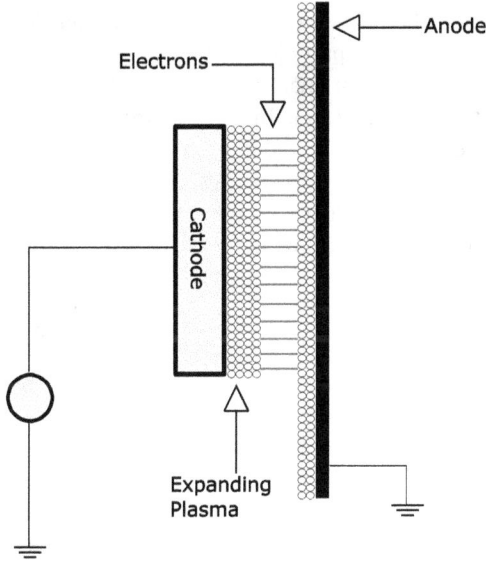

Figure 7.2   High-power electron beam diode

The electrons in the plasma are accelerated by the diode electric forces that are produced by the applied voltage. The charge in the diodes produces, as well, a new electric field as described by the Child-Langmuir (CL) law (see (7.1)). New electrons cannot be accelerated when the resultant electric field between the produced and the applied is zero in the plasma vicinity. This phenomenon is known as space charge limited current [8–12]. The maximum current density for the infinitely long parallel plates can be stated as follows:

$$J_{1DCL} = \frac{4}{9} \varepsilon_0 \sqrt{\frac{2e}{m}} \frac{V^{3/2}}{d^2}, \tag{7.1}$$

where ε_0 is the free space permitivity, m is the rest electron mass, e is the electron charge, d is the gap or distance between electrodes and V the voltage applied.

As mentioned earlier, two phenomena are occurring in this region that limit the performance of the Vircator. The first one, known as gap closure [13–15], restricts the operation time. The second one is electron divergence, which study is the main goal of this chapter and will be further explained in Section 7.1.3.

7.1.2   Drift region

The anode is the interface between the Diode and the Drift regions. The anode sets the potential to accelerate the particles, allowing them to pass through it. The main characteristic of the anode is its transparency (T_a), which defines the percentage of particles that pass through it.

In the Drift region, the particles travel free of external forces. However, each particle interacts with the other particles in the beam and the boundary conditions. This situation defines a limit or maximum current that can drift. This is called space-charge limiting current (I_{SCL}). In 1961, Birdsall *et al.* noticed that if the current injected in the drift space exceeds I_{SCL}, the charge accumulates in a region of the space producing a Virtual Cathode (VC) [16–18].

Microwave radiation is produced by two phenomena. The former refers to the electron oscillation between the real cathode and the VC, while the latter pertains to the VC oscillation itself.

7.1.3 Beam divergence

The motion of electrons in the diode region is influenced by three forces: self-electric, self-magnetic, and electrostatic. Numerous studies have explored this subject [19–21]. The self-electric force arises due to the electric charge of the particles, which causes repulsion between them. This force results in the divergence of the electron beam. In contrast, the self-magnetic force causes the beam to pinch, with the resulting force being significant for high-energy beams. The electrostatic force is determined by the shape of the electrodes as well as the applied voltage determining the initial paths of the electrons. These three forces cause beam dispersion or beam pinching. Vircator power generation capability strongly depends on the electron beam injected in the drift-tube region [22–24] being optimal when these three forces produce a one-dimensional beam [5]. Figure 7.3 presents a schematic of a Vircator where the diode current, beam current, diverged current, and divergence angle are shown.

Beam divergence can have a significant impact on the performance of the Vircator. For example, in [25], the author reports losses of up to 30% due to

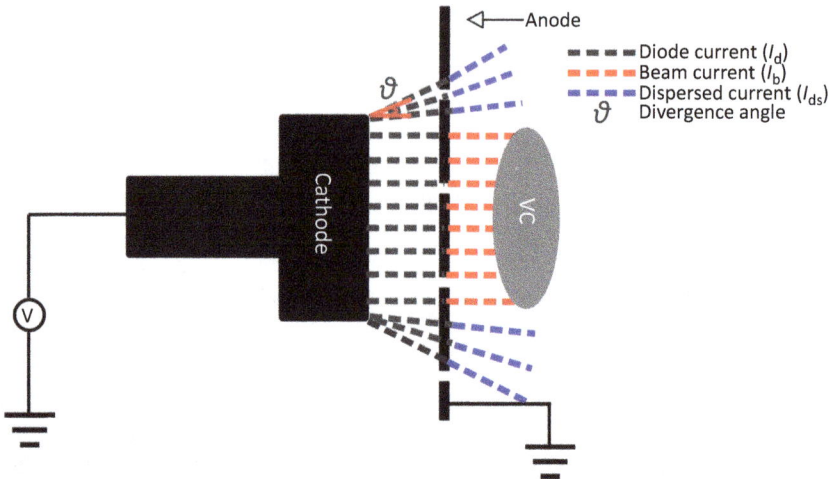

Figure 7.3   Beam divergence schematic

electron divergence. On the other hand, for a planar circular cathode with a straight corner, the divergence disappears, for functional Vircators, at relativistic voltages ($V > 511$kV) [20] and a high magnitude of diode current as

$$I_c = \frac{2\pi\varepsilon_0 mc^3}{e} \frac{r_c}{d} \sqrt{\gamma_0^2 - 1}, \qquad (7.2)$$

where c is the light speed, r_c the cathode radius and γ_0 is the relativistic factor given by $\gamma_0 = eV/mc^2 + 1$.

Equation (7.2) shows that to achieve one-dimensional flow at lower voltages and currents, electrodes should be profiled in order to control the electrostatic force and reduce electron divergence. However, finding analytical solutions for specific electrode profiles can be cumbersome. In this case, numerical and experimental approaches are more suitable. For example, in [26], several cathode profiles were tested experimentally for a specific application at a given voltage. The following section introduces some simulation tools and presents one example of a cathode profiling design aimed at reducing beam divergence.

7.2 Numerical analysis of beam divergence in vircators

In this section, we will showcase an example of employing numerical simulations for designing a diode that operates under non-relativistic voltage conditions while simultaneously minimizing beam divergence.

The study of cathode profiling can be carried out using Particle-in-Cell (PIC) simulation software. This numerical approach entails solving the governing equations for magnetic and electric fields, represented by Maxwell's equations, as well as the motion of charged particles, described by the Lorentz force equation, in the time domain [27,28]. Among the commercial software or codes that can be used are Computer Simulation Technology Particle Studio (CST-PS), X11-based Object-Oriented Particle-In-Cell (XOOPIC), and MAGIC.

These tools can be employed to investigate various methods for mitigating beam divergence, such as incorporating a rim into the cathode structure. This study will thoroughly examine the impact of a rim on the emission regions while maintaining the essential electrostatic conditions, as illustrated in Figure 7.4. This technique shifts the particles away from the border region where the radial component is high, ultimately reducing the impact of this high-field zone on the emitted beam.

7.2.1 Methodology

The methodology was carried out in two steps.

1. Validation of the simulations: In this step, the simulation results are compared with well-known benchmarks. For this example, the two-dimensional correction of the Child–Langmuir law, deduced by R. Kelly [29], was

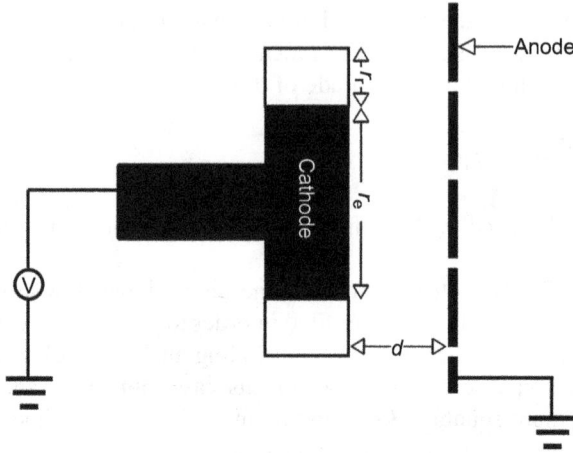

Figure 7.4   Cathode with rims. The scheme shows the emitting radius r_e, the rim radius r_r, and the gap separation length d.

validated in XOOPIC [28]. R. Kelly concludes that the ratio of the J_{2DCL} to the theoretical 1D value J_{1DCL} is always decreasing function of the ratio of emission radius to gap separation. The derived prediction from reference [29] is shown in (7.3).

$$\frac{J_{2DCL}}{J_{1DCL}} \approx 1 + 0.419\frac{d}{r_c} + 0.036\left(\frac{d}{r_c}\right)^2. \tag{7.3}$$

2.   Parametric Variation: In this step, the effects of the rim radius (r_r) variation for a constant emission radius of 30 mm, voltage (V), and gap (d) are studied. The rim radius was varied from 2 mm to 14 mm with increments of 2 mm. See Figure 7.4 for reference.

7.2.2   Results

Figure 7.5 shows the results of validating the simulation step. On the plot, the theoretical (Equation (7.3)) and simulated ratio J_{2DCL}/J_{1DCL} is shown. This step ensures that the simulation setup for the given condition is consistent with the theoretical framework, validating the accuracy of the simulation results.

Figure 7.6 presents a lateral view of two simulations performed during the parametric variation step. A cathode without a rim, presented in Figure 7.6(a), exhibits a higher dispersion of electrons in comparison with the cathode with a rim, which is presented in Figure 7.6(b). This means that a substantial amount of electrons are emitted from the cathode brim due to the edge effect, which leads to an increase in electron losses.

Further analysis is done to evaluate the emitted current by cathodes with different rim radii (see Figure 7.7). As the rim size ratio increases, the current decreases. The presence of a rim around the cathode reduces the edge effect, which

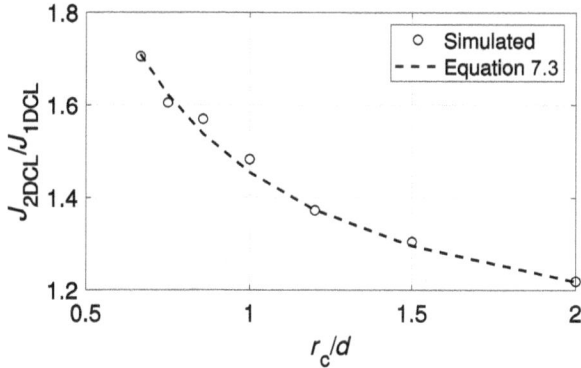

Figure 7.5   Results of the validation step. Comparison between J_{2DCL}/J_{1DCL} theoretical and XOOPIC simulations.

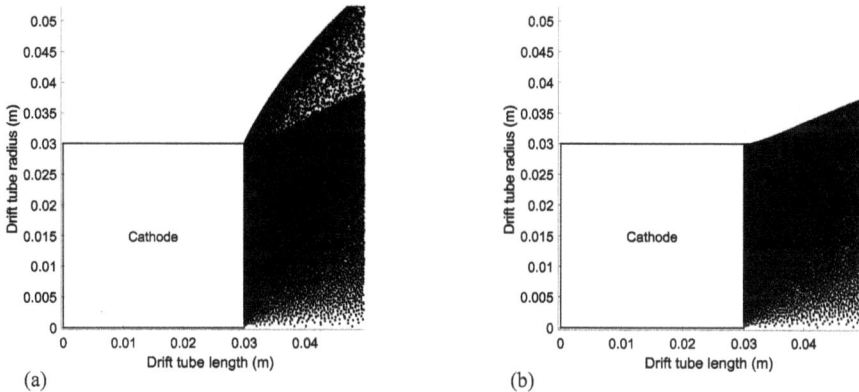

Figure 7.6   Lateral view of the cathodes XOOPIC simulation. (a) Without rim and (b) with rim of 2 mm.

is the tendency of electrons to be emitted from the brim of the cathode. Without the rim, a significant amount of electrons are emitted from the edges, leading to electron losses and higher dispersion. However, when a rim is present, it shields the brim of the cathode and reduces the edge effect, leading to a lower dispersion (see Figure 7.8) of electrons and a lower current being emitted. The dispersion of electrons for a cathode without rim is around 31° compared to a cathode with a rim, where the dispersion was only 2° to 4°. The inclusion of a rim in the cathode design contributes to a reduction in electron dispersion, resulting in lower current output compared to cathodes without rims. However, despite the decrease in current, cathodes with rims may still be advantageous in applications where energy efficiency is of paramount importance.

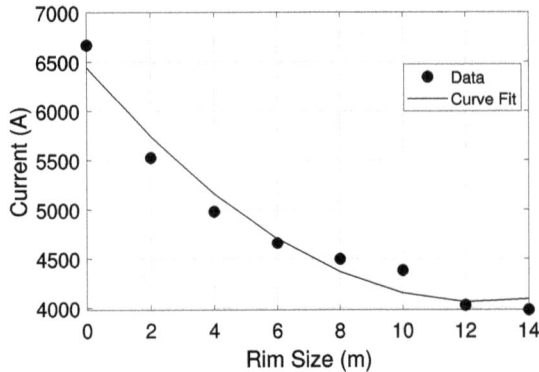

Figure 7.7   Variation of emitted current for a diode with a 20 mm gap, 30 mm emitting radius, and 350 kV voltage as the rim radius varies parametrically

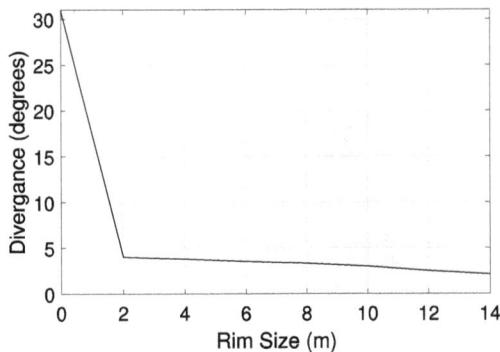

Figure 7.8   Variation of divergence angle for a diode with a 20 mm gap, 30 mm emitting radius, and 350 kV voltage as the rim radius varies parametrically

The next section will overview of the experimental techniques measuring electron dispersion and beam homogeneity.

7.3   Experimental analysis of beam divergence

Analyzing the beam divergence and homogeneity is important in developing and optimizing high-power microwave devices. Owing to a better understanding of the beam characteristics can lead to improvements in device efficiency, power output, and reliability.

Experimental measurement of beam divergence and homogeneity of electrodes can be achieved using various methods [30–33]. The most common methods used for this purpose are beam imprints on radiochromic films, pepper-pot method, cameras, and spectrometers. In this chapter, full beam imprints on radiochromic films [34] have been used to analyze the homogeneity. Additionally, the pepper-pot method has been employed to obtain transverse emittance, which, in turn, provides the angle of divergence. It is important to note that this method was utilized by many researchers in the past [26,35–37]. Although the method is relatively old it can still provide valuable information. However, it is important to note certain limitations inherent to this method, including:

- Time-integrated measurement: The radiochromic film measures the cumulative radiation dose they are exposed to over time. They do not capture the instantaneous intensity or changes in beam divergence that occur at a specific instance of time. This makes it unsuitable for real-time monitoring [38].
- Dose-rate dependency: The radiochromic film response can vary depending on the radiation they are exposed to. This can lead to inaccuracies during beam divergence measurement, especially with beams that have rapid fluctuations.
- Limitation of spatial resolutions: Although the resolution of the radiochromic films is good compared to other dosimeter films they are still limited in capturing fine details in the beam profile, especially beams with small sizes. This affects the divergence measurement accuracy [39].

Currently, advanced spectrometers and cameras are being utilized. Most notable works include the research paper by M. Barbisan [40] where beam emission spectroscopy diagnostic was utilized to measure the uniformity and the divergence of the beam. However, in this chapter, the radiochromic film technique is used because it offers a simpler and more accessible approach to measuring beam homogeneity compared to advanced spectroscopy and cameras. This simplicity is advantageous in settings where resources or specialized equipment are limited, making the method more inclusive and widely applicable.

7.3.1   Methodology

In order to evaluate the beam homogeneity characteristics and transverse emittance profiles, two planar cathodes were investigated. The first one is graphite-based, whereas the second is a carbon epoxy multi-capillary. A detailed description of the examined cathodes can be found in Table 7.1.

During the experiment, the cathodes were mounted on a holder. The anode was a stainless steel mesh with a transparency of 77%. The diode region was vacuumed up to $\sim 5 \times 10^{-5}$ mbars. Figure 7.9 shows the experimental setup. The A-K gap distance d was 20 mm, and the distance between the anode and collimator (see Figure. 7.10), D was 3 mm. To calibrate the films, a nanosecond pulsed accelerator was used [41]. The accelerator provides electron beam energies exceeding 10 MeV, with a beam current bigger than 1 A, pulse duration of about 7 μs, and pulse repetition rates of 400 Hz [26].

In the subsequent step, two methods, one for acquiring the beam homogeneity information and the other for obtaining beam divergence, are explained.

Table 7.1   Examined cathode emitter material, emitter diameter, total diameter (emitter with screening ring), and description

Cathode number	Emitter	D_1 [mm]	D_2 [mm]	Description
1	MPG-8 graphite	50	60	Smooth planar surface
2	Carbon epoxy multi-capillary cathode	70	90	Smooth planar surface substrate with capillaries attached

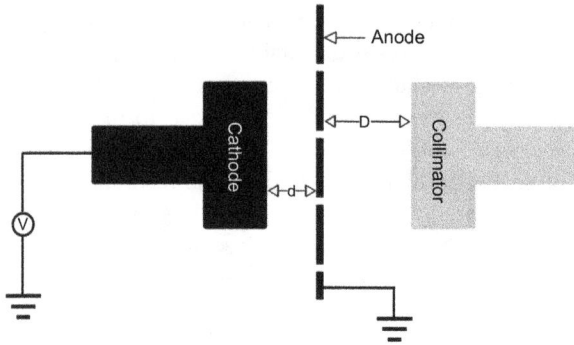

Figure 7.9   Schematic of the experimental setup

Figure 7.10   Collimator used during the experiments. Holes about 2 mm in diameter with gaps between them of 5 mm.

1. **Beam Homogeneity:** The following is a step-by-step description of the experiment focusing on getting the full beam imprint. The radiochromic films are placed on top of the collimator and radiated with electron beams. The images captured from irradiated radiochromic films are obtained using a transmission-mode scanner (Thru-Scan) [42], where the scanner's exposure time is held constant. Scanning the films with known absorbance values can determine the beam profile by correlating the magnitude of the pixels and the

absorbed dose. The absorbed dose quantifies the energy deposited by radiation per unit mass. The beam profile obtained provides insights into the electron beam's homogeneity. The use of radiochromic films offers a non-destructive method for evaluating beam uniformity.

A calibration curve is generated to establish the relationship between pixel values and absorbed dose. To do this, radiochromic films are exposed to known radiation doses, and the resulting pixel values are measured using the scanner. This procedure is repeated for various doses. Once the curve is established, absorbed doses can be determined from the pixel values obtained from the scanned films.

Table 7.2 presents the results of measuring optical density and absorbed dose for samples using radiochromic films for cathode #1 and cathode #2:

In Table 7.2, the first column presents the Optical density (or optical absorbance), which measures the amount of light absorbed by a material at a specific wavelength. In this case, the optical density was measured at a wavelength of 512 nm. In the second column, the absorbed dose quantifies the energy deposited by radiation per unit mass of the irradiated material.

2. **Beam Divergence:** To measure the transverse emittance (divergence of the beam), the film is positioned behind the collimator applying the pepper-pot method [35]. The pepper-pot method is a technique that provides data about the beam profile and angular displacement. This method is not susceptible to space-charge effects, making it suitable for Vircator measurement applications. A single measurement can yield vertical and horizontal emittance values in an experiment.

In the pepper-pot method, the incident beam is divided into smaller beamlets. Information about the beam profile and angular divergence is obtained by measuring the spot centroid's displacement from the corresponding aperture.

When the beam interacts with the pepper-pot plate, the main beam component is blocked, allowing only small beamlets to pass through perforations and reach the screen. The image produced on the screen is detected and analyzed, enabling the determination of angular displacement (x') in terms of a slope from point (x, y) at the pepper-pot plate to point (u, v) on the

Table 7.2   Results of measuring the optical density and absorbed dose of a sample using radiochromic film for cathode #1 and cathode #2

Optical density	Absorbed dose [kGy]	Sample number
0.568	32.38	1
1.077	64.56	2
1.547	96.41	3
1.959	128.2	4

screen, see (7.4).

$$x' = \left(\frac{u-x}{L}\right), \quad y' = \left(\frac{u-y}{L}\right), \tag{7.4}$$

where L is the distance between the pepper-pot plate and the screen.

The employed image processing technique is analogous to that used for full beam imprints. Figure 7.11 presents a schematic of the typical setup for transverse emittance measurement using the pepper-pot method.

The beam's converging or diverging nature can be determined by measuring the displacement of the spot centroid from the corresponding aperture.

In summary, the full beam imprint analysis and transverse emittance measurement provide insights into the electron beam's homogeneity and angular characteristics. These non-destructive techniques facilitate the evaluation and optimization of charged particle devices, improving performance and reliability. Figure 7.12 summarizes both methods in a flowchart.

7.3.2   Results

The results section is divided into two parts: one for measuring beam homogeneity and another for analyzing beam divergence.

1. **Beam homogeneity:** Figure 7.13 displays the results of the scanning step. The radiochromic films on both cathodes received equal exposure times, resulting in comparable doses and mean pixel values. The calibration curve, which illustrates the correlation between doses and pixel values for both cathodes, is presented in Figure 7.14. The complete beam imprints for cathode #1 and cathode #2 can be observed in Figures 7.15 and 7.16, respectively.

    The beam imprints provide information about cathode uniformity. Examining the beam imprint for cathode #1 (Figure 7.15) reveals a non-uniform current

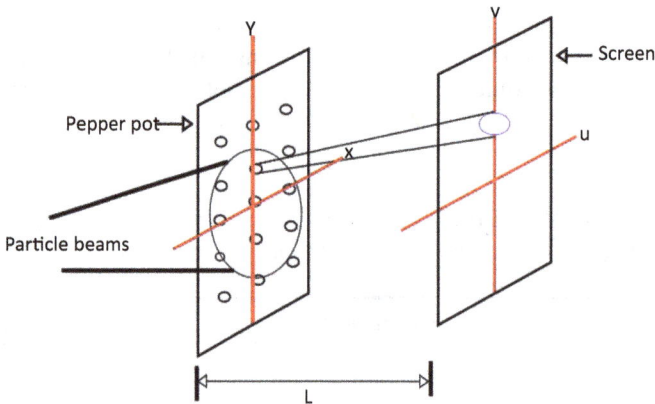

Figure 7.11   Schematic for the pepper-pot method

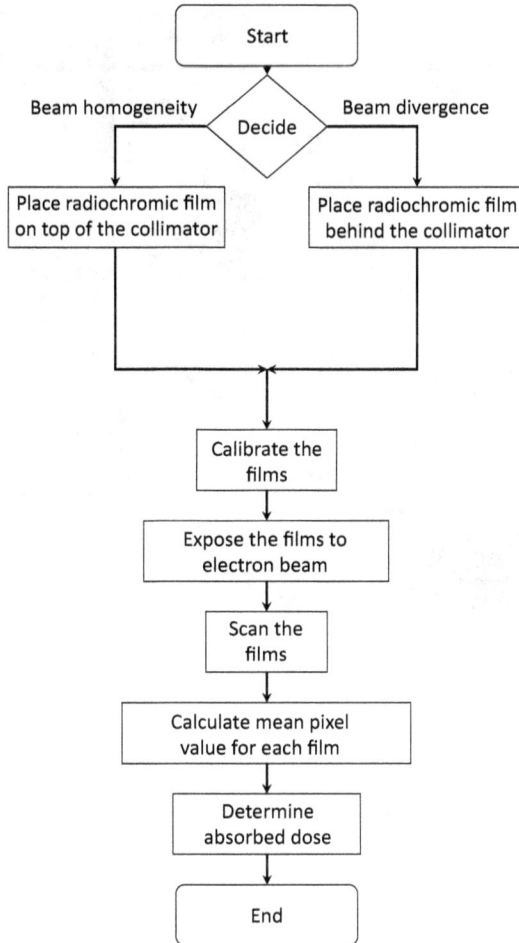

Figure 7.12 Flowchart illustrating the image processing steps for beam imprint acquisition and the decision between analyzing beam homogeneity or implementing the Pepper-Pot method for measuring beam divergence

density distribution, which may be due to the edge effect. The edge effect occurs when the field strength at the periphery of the cathode exceeds that at the center. This phenomenon is supported by the presence of a discernible white spot in the beam imprint.

To fully understand the homogeneity of the cathodes, it is necessary to consider the behavior of electrons within the plasma. Electrons are confined by strong magnetic fields and do not escape the plasma easily. This confinement contributes to the overall stability of the plasma, which further impacts the homogeneity of the cathodes.

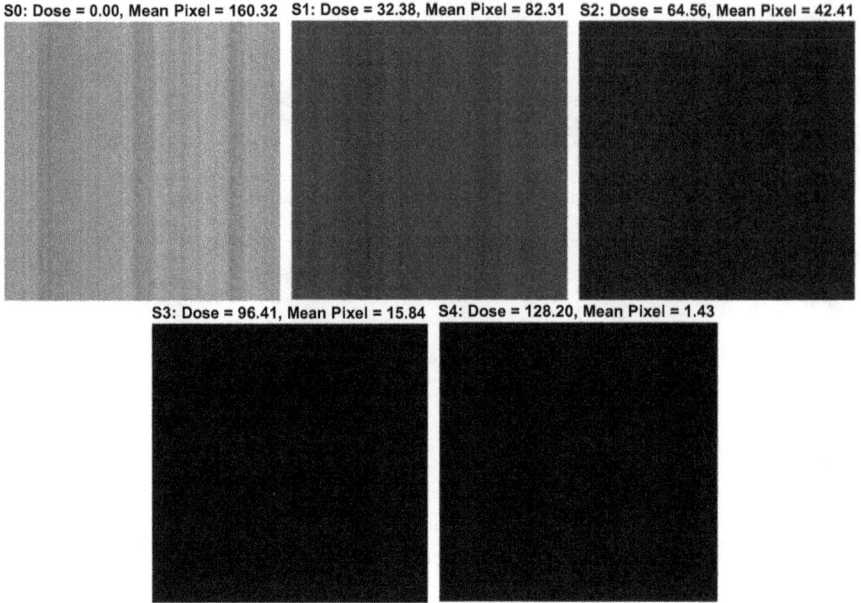

Figure 7.13   Radiochromic film radiation doses and mean pixel value

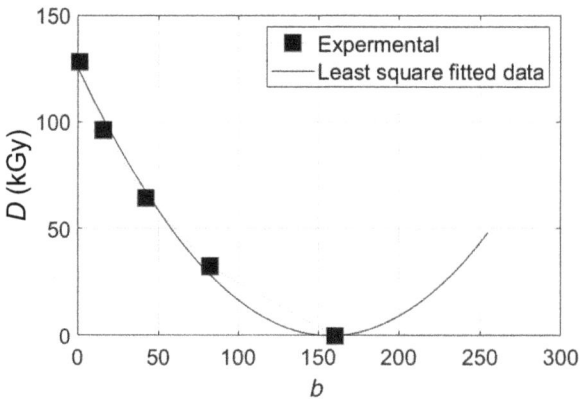

Figure 7.14   Absorbed dose versus pixel for cathode #1 and cathode #2

Upon examination of the full beam imprint of cathode #2, a more homogeneous distribution of current density is apparent compared to cathode #1. This improved uniformity can be attributed to the presence of carbon epoxy multi-capillaries, which promote even distribution of current density. Unlike cathode #1, there are no observable density dips in the center, and the current density does not exhibit stronger emission at the edges.

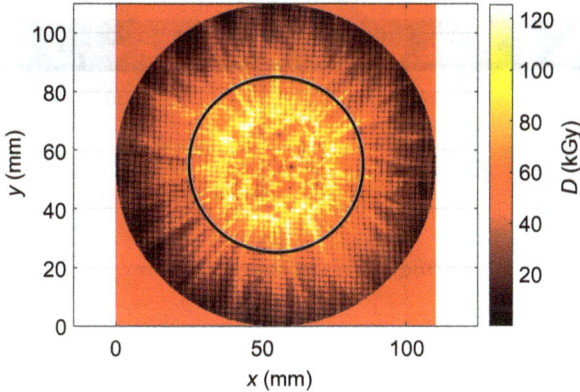

Figure 7.15   Figure shows beam imprint for cathode #1

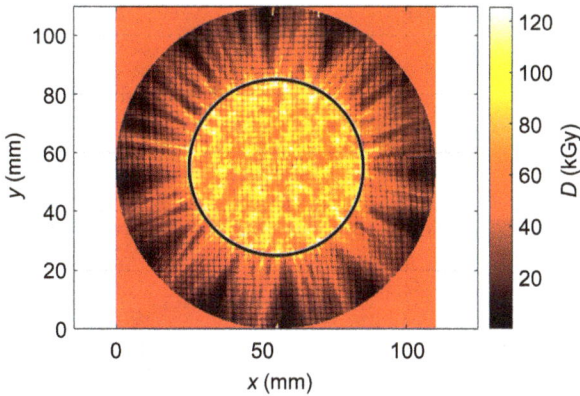

Figure 7.16   Figure shows beam imprint for cathode #2

The inclusion of carbon epoxy multi-capillaries in cathode #2 demonstrates their effectiveness in mitigating the edge effect, resulting in a more stable and uniform plasma. This improvement underscores the significance of utilizing advanced materials and design strategies in cathode development, as they can ultimately optimize performance and enhance control over the plasma generation process.

2. **Beam divergence:** To assess the beam divergence, digital images were analyzed to determine the displacement of the beamlets in relation to their corresponding hole positions on the pepper-pot screen. Figure 7.17 illustrates the horizontal beam imprints obtained on the radiochromic film situated behind the collimator. These imprints were used to calculate the displacement of the microbeam centers relative to the centers of the collimating holes as a function of the hole position, employing spatial calibration and image processing techniques.

The characteristic angles of electron incidence on the anode surface can be calculated based on the magnitudes of displacements and the thickness of the collimator

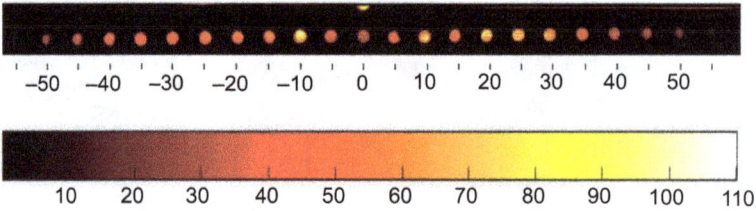

Figure 7.17   Figure shows the beam imprint for cathode #1 placed behind the collimator

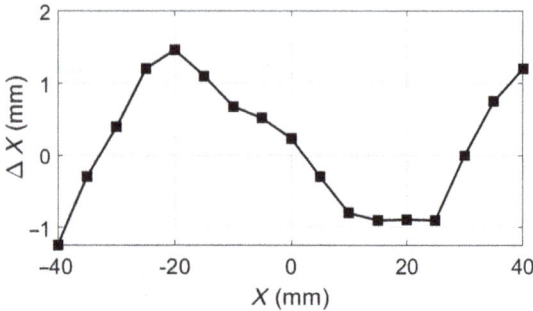

Figure 7.18   Displacement of the microbeams' centers relative to the centers of the collimating holes versus the hole position for cathode #1

(8 mm). Figure 7.18 displays the displacement of the microbeam centers relative to the centers of the collimating holes as a function of the hole position for cathode #1.

The calculated characteristic angles range from 0 to 10.6473°, which indicates the variation in electron incidence on the anode surface resulting from beam divergence.

The perpendicular momentum, $p_\perp$, quantifies the deviation of individual electrons from the primary direction of the beam. This quantity is obtained by projecting the total momentum, p_{total}, onto a plane that is perpendicular to the primary direction of the beam. The angle between the total momentum and the primary direction of the beam is utilized in this projection, as depicted in (7.5).

$$p_\perp = p_{total} \cdot \sin(\theta), \tag{7.5}$$

where θ is the angle between the total momentum and the main direction of the beam. The larger the angle, the larger the perpendicular momentum.

The spread is then calculated by comparing the perpendicular momentum to the total momentum. This is done by dividing $p_\perp$ by p_{total} and expressing the result as a percentage (see (7.5))

$$\text{Spread} = \frac{p_\perp}{p_{total}} \times 100\% = 18.7\%. \tag{7.6}$$

The perpendicular momentum is approximately 2.22×10^{-23} kg m/s, and the spread is about 18.7%. In conclusion, the beam divergence of an electron beam is an essential characteristic that can impact the beam's behavior as it propagates through space. The analysis performed in this section has revealed a relatively high beam divergence of 18.7%, which could affect the electron beam's effectiveness and precision in certain applications. By calculating the perpendicular momentum and the spread for various angles, it is possible to quantify the divergence and better understand the electron beam's properties. The insights gained into the relationship between the displacement of microbeams' centers relative to the centers of the collimating holes and the angles of incidence on the anode surface can be used to optimize the design and performance of cathodes.

Further research may be necessary to develop strategies to minimize beam divergence and enhance the overall performance of cathodes.

7.4   Conclusions

This chapter focuses on both numerical and experimental analysis of beam divergence in vircators. The numerical analysis using Particle-in-Cell (PIC) simulation software examines cathode profiling and the incorporation of a rim into the cathode structure to mitigate beam divergence. The results show a notable reduction in electron dispersion as the rim size is increased, leading to improved performance and efficiency of the electron beam.

Moreover, this chapter also delves into the topic of beam homogeneity, a critical factor in the performance of high-power microwave devices. Beam homogeneity, or the consistency of the beam's intensity across its cross-section, plays a significant role in the device's efficiency and output power. For the analysis of beam homogeneity, radiochromic films are employed. These are radiation-sensitive films that undergo a color change upon exposure to a beam. This allows for the recording of the beam's spatial distribution and the observation of any variations in intensity across its cross-section.

Furthermore, the chapter explores the experimental analysis of beam divergence using the pepper-pot method, which is employed to obtain transverse emittance imprints and analyze beam homogeneity in conjunction with radiochromic films. This experimental approach provides valuable insights into the behavior of electron beams in high-power microwave devices and allows for a comprehensive understanding of the beam characteristics.

By integrating the findings from both numerical and experimental analyses, this chapter contributes to the optimization of cathodes and vircators, with the ultimate goal of improving the performance and reliability of these devices in their respective applications. This comprehensive approach to understanding beam divergence can pave the way for the development of better diagnostic tools and techniques in the field of high-power microwave devices.

References

[1]   Benford J, Swegle JA, and Schamiloglu E. *High Power Microwaves*. 2nd ed. Boca Raton, FL: Taylor & Francis; 2007.

[2]   Scott K. *Coaxial Vircator Geometries*. Texas Tech University; 1998.

[3]   Walter J, J Vara and CL, Dickens J, *et al.* Initial anode optimization for a compact sealed tube vircator. In: 2011 IEEE Pulsed Power Conference; 2011. p. 807–810.

[4]   Parson JM, Mankowski JJ, Dickens JC, *et al.* Imaging of explosive emission cathode and anode plasma in a vacuum-sealed vircator high-power microwave source at 250 A/cm^2. *IEEE Transactions on Plasma Science*. 2014;42 (10):2592–2593.

[5]   Neira E. *Study on the Optimization of Virtual Cathode Oscillators for High Power Microwaves Testing*. Universidad Nacional de Colombia. Bogota, Colombia; 2020. Available from: https://repositorio.unal.edu.co/handle/unal/75948.

[6]   Litvinov EA, Mesyats GA, and Proskurovskia DI. Field emission and explosive electron emission processes in vacuum discharges. *Soviet Physics Uspekhi*. 1983;26(2):138. Available from: http://stacks.iop.org/0038-5670/ 26/i=2/a=R03.

[7]   Mesyats GA. Vacuum discharge effects in the diodes of high-current electron accelerators. *IEEE Transactions on Plasma Science*. 1991;19(5):683–689.

[8]   Child CD. Discharge from hot CaO. *Physical Review (Series I)*. 1911; 32:492–511. Available from: http://link.aps.org/doi/10.1103/PhysRevSer-iesI.32.492.

[9]   Langmuir I. The effect of space charge and residual gases on thermionic currents in high vacuum. *Physical Review*. 1913;2:450–486. Available from: http://link.aps.org/doi/10.1103/PhysRev.2.450.

[10]   Langmuir I, and Blodgett KB. Currents limited by space charge between coaxial cylinders. *Physical Review*. 1923;22:347–356. Available from: http:// link.aps.org/doi/10.1103/PhysRev.22.347.

[11]   Carr CG. *Space Charge Limited Emission Studies Using Coulomb's Law*. Naval Postgraduate School. Monterrey, California; 2004. Available from: https://calhoun.nps.edu/handle/10945/1175.

[12]   Zhang P, Valfells G, Ang LK, *et al.* 100 years of the physics of diodes. *Applied Physics Reviews*. 2017;4(1):011304. Available from: https://doi.org/ 10.1063/1.4978231.

[13]   Petren J. Frequency tunability of axial cavity vircators and double anode vircators. *KTH, School of Electrical Engineering (EES)*; 2016.

[14]   Coleman JE, Moir DC, Ekdahl CA, *et al.* Explosive emission and gap closure from a relativistic electron beam diode. In: 2013 *19th IEEE Pulsed Power Conference (PPC)*; 2013. p. 1–6.

[15]   Roy A, Menon R, Mitra S, *et al.* Plasma expansion and fast gap closure in a high power electron beam diode. *Physics of Plasmas*. 2009;16(5):053103. Available from: https://doi.org/10.1063/1.3129802.

[16] Sullivan DJ. Applications of the virtual cathode in relativistic electron beams. In: *1979 3rd International Topical Conference on High-Power Electron and Ion Beam Research & Technology*. vol. 2; 1979. p. 769–772.

[17] Sullivan DJ. A high frequency vircator microwave generator. *High-Power Particle Beams*. 1983;p. 557–560.

[18] Birdsall CK, and Brides WB. Space-charge instabilities in electron diodes and plasma convertes. *Journal of Applied Physics*. 1961;32:2611–2618.

[19] Hramov AE, Kurkin SA, Koronovskii AA, *et al.* Effect of self-magnetic fields on the nonlinear dynamics of relativistic electron beam with virtual cathode. *Physics of Plasmas*. 2012;19(11):112101. Available from: https://doi.org/10.1063/1.4765062.

[20] Putnam S. *Theoretical Studies of Intense Relativistic Electron Beam-Plasma Interactions*. 1st ed. San Leandro, CA: Physics International Company; 1972.

[21] Woo W. Two-dimensional features of virtual cathode and microwave emission. *The Physics of Fluids*. 1987;30(1):239–244. Available from: https://aip.scitation.org/doi/abs/10.1063/1.866181.

[22] Alyokhin BV, Dubinov AE, Selemir VD, *et al.* Theoretical and experimental studies of virtual cathode microwave devices. *IEEE Transactions on Plasma Science*. 1994;22(5):945–959.

[23] Biswas D, and Kumar R. Microwave power enhancement in the simulation of a resonant coaxial vircator. *IEEE Transactions on Plasma Science*. 2010; 38(6):1313–1317.

[24] Neira E, Xie YZ, and Vega F. On the virtual cathode oscillator's energy optimization. *AIP Advances*. 2018;8(12):125210. Available from: https://doi.org/10.1063/1.5045587.

[25] Choi EH, Choi MC, Jung Y, *et al.* High-power microwave generation from an axially extracted virtual cathode oscillator. *IEEE Transactions on Plasma Science*. 2000;28(6):2128–2134.

[26] Baryshevsky V, Gurinovich A, Gurnevich E, *et al.* Experimental study of a triode reflex geometry vircator. *IEEE Transactions on Plasma Science*. 2017; 45(4):631–635.

[27] Verboncoeur JP, Alves MV, Vahedi V, *et al.* Simultaneous potential and circuital solution for 1D Bounded plasma particle simulation codes. *Computational Physics*. 1993;104(2):321–328.

[28] Verboncoeur JP. OOPIC: object oriented particle-in-cell code. In: *International Conference on Plasma Science (papers in summary form only received)*; 1995. p. 244.

[29] Ragan-Kelley B, and Verboncoeur J. Two-dimensional axisymmetric Child–Langmuir scaling law. In: *2008 IEEE International Vacuum Electronics Conference*; 2008. p. 211–211.

[30] Thomas RL, Nusinovich GS, Granatstein VL, *et al.* Overview of high-power microwave sources and applications. *IEEE Transactions on Plasma Science*. 2021;49(1):274–301.

[31] Brunetti E, Shanks R, Manahan G, *et al.* A novel diagnostic for measuring the transverse energy spread of electron beams from laser-plasma

accelerators. In: *2010 IEEE International Conference on Plasma Science (ICOPS)*. Piscataway, NJ: IEEE; 2010. p. 1–1.

[32]   Andersen K, and Andersen H. A novel method for the measurement of the transverse beam quality factor. *Nuclear Instruments and Methods in Physics Research Section A: Accelerators, Spectrometers, Detectors and Associated Equipment.* 2004;528(1–2):101–108.

[33]   Mehrvarz H, Doury B, Beaudoin B, *et al.* Measurement of the transverse energy spread in a high-current relativistic electron beam. *Nuclear Instruments and Methods in Physics Research Section A: Accelerators, Spectrometers, Detectors and Associated Equipment.* 2013;728:121–125.

[34]   Mukherjee B, Makowski D, Krasinski P, *et al.* Novel applications of radiochromic film in radiation dosimetry at high-energy accelerators. In: *2008 15th International Conference on Mixed Design of Integrated Circuits and Systems*; 2008. p. 131–134.

[35]   Wang JG, Wang DX, and Reiser M. Beam emittance measurement by the pepper-pot method. *Nuclear Instruments and Methods in Physics Research Section A: Accelerators, Spectrometers, Detectors and Associated Equipment.* 1991;307(2–3):190–194. Available from: http://dx.doi.org/10.1016/0168-9002(91)90182-P.

[36]   Nürnberg F, Schollmeier M, Brambrink E, *et al.* Radiochromic film imaging spectroscopy of laser-accelerated proton beams. *Review of Scientific Instruments.* 2009;80(3):033301. Available from: http://dx.doi.org/10.1063/1.3086424.

[37]   Gonzalez-Lopez A, Lago-Martin JD, and Vera-Sanchez JA. Small fields measurements with radiochromic films. *Journal of Medical Physics.* 2015;40(2):61. Available from: http://dx.doi.org/10.4103/0971-6203.158667.

[38]   Dunn L, Godwin G, Hellyer J, *et al.* A method for time-independent film dosimetry: Can we obtain accurate patient-specific QA results at any time postirradiation? *Journal of Applied Clinical Medical Physics.* 2022;23(3). Available from: http://dx.doi.org/10.1002/acm2.13534.

[39]   Miyatake T, Kojima S, Sakaki H, *et al.* Evaluation of the spatial resolution of Gafchromic^{TM} HD-V2 radiochromic film characterized by the modulation transfer function. *AIP Advances.* 2023;13(8). Available from: http://dx.doi.org/10.1063/5.0160754.

[40]   Barbisan M, Bonomo F, Fantz U, *et al.* Beam characterization by means of emission spectroscopy in the ELISE test facility. *Plasma Physics and Controlled Fusion.* 2017;59(5):055017. Available from: http://dx.doi.org/10.1088/1361-6587/aa6584.

[41]   Korenev S, and Korenev I. Nanosecond pulsed electron accelerators for radiation technologies. In: *Conference Record of the Twenty-Sixth International Power Modulator Symposium, 2004 and 2004 High-Voltage Workshop*; 2004. p. 278–281.

[42]   Méndez I, Rovira-Escutia JJ, and Casar B. A protocol for accurate radiochromic film dosimetry using Radiochromic.com. *Radiology and Oncology.* 2021;55(3):369–378.

Chapter 8

Use of C-UAS system and its EM effect analysis

Tae Heon Jang[1], Jeong Ju Bang[2] and Jeong Min Kim[2]

Unmanned aerial vehicles (UAVs), commonly known as drones, were initially developed for military use, but have recently emerged as a new mobility platform that combines ICT and AI technologies. With the development of related technologies such as GPS, navigation devices, and batteries, high-performance UAVs are becoming increasingly popular. Private applications of UAVs are growing rapidly, including real-time monitoring, wireless coverage delivery, remote sensing, search and rescue, commodity delivery, security and surveillance, precision agriculture, and private infrastructure inspection. Several security incidents involving UAVs have been witnessed around the world, making them a real threat. To address this issue, many countries have begun implementing counter-UAS (C-UAS) systems in airports (civil and military) and major facilities, including UAS detection and/or neutralization. A C-UAS system is a set of technical tools that can monitor, detect, identify, record, and respond to unauthorized UAS (unmanned aircraft system) activities. C-UAS systems may also include measures to neutralize or limit potential risks, and can be deployed in fixed positions, mounted on vehicles or drones, or made portable. Most C-UAS systems use radar and radio frequency (RF) jammers for detecting and neutralizing UAS. However, due to their relatively high RF output, they can affect existing critical systems, subsystems, or peripheral equipment. This chapter examines the technology trends of UAS and C-UAS systems and explores the electromagnetic characteristics of C-UAS systems. The chapter also looks at IEC standards and other standards for assessing electromagnetic (EM) vulnerability in national critical infrastructures caused by EM or high power electromagnetic (HPEM) sources in terms of HPEM threats to mission drones and robots.

8.1 Application of UAV technology and threat

Innovation of UAVs has been rapidly advanced due to the development of technologies of lightweight materials, sensors, communications, batteries, artificial intelligence, propulsion systems, and software.

[1]Standard and Certificate Division, Space & Bean, Korea
[2]Aerospace EM Technology Center, Korea Testing Laboratory, Korea

8.1.1   Definition of UAVs

Depending on the purpose of use, institution, and timing of use, UAV is used under various names as follows: ultra-light aircraft, drones, remotely piloted vehicles (RPV), UAV, UAS, RPA (remotely piloted aircraft), etc. UAV is defined in various ways according to countries, organizations, and institutions [1].

Definition of the term "unmanned aerial vehicle" per official documentation of the United States Department of Defense is a powered, aerial vehicle that does not carry a human operator, uses aerodynamic forces to provide vehicle lift, can fly autonomously or be piloted remotely, can be expendable or recoverable, and can carry a lethal or non-lethal payload [2].

In addition, the Federal Aviation Administration (FAA) defines unmanned aerial vehicles as aircraft, including all types of airplanes, helicopters, and airships that are used for aerial flight purposes without pilots on board [3].

8.1.2   UAV application area

Unmanned aircraft systems, so-called UAVs, or so-called drones, are among the major growing technologies that have many beneficial applications. The recreational and commercial use of UAS is on a steep increase. UAS is employed by civilian engineers for seismic risk assessments, transportation planning, disaster response, and construction management. Even in the nuclear industry, new uses for UAS to fulfill operational, safety, and environmental monitoring tasks are continually being explored, including taking physical, chemical, and radiochemical measurements; extending human safety capabilities by monitoring in environments where humans cannot go; and, expanding the deployment of traditional security detection (e.g., sensors and cameras) and perimeter monitoring systems [4].

8.1.3   UAV technology trends

The use of lightweight materials, such as carbon fiber and titanium, has allowed for the construction of smaller and lighter UAVs that are more maneuverable and efficient. The development of advanced sensors, such as GPS, inertial navigation systems, and cameras, has enabled UAVs to be flown autonomously, accurately, and safely. The use of reliable wireless communication systems has allowed UAVs to be controlled and monitored from a distance, enabling their use in remote or dangerous environments. The development of high-capacity, lightweight batteries has increased the endurance and range of UAVs, making them more practical for a variety of applications. The integration of artificial intelligence and machine learning has enabled UAVs to perform more complex tasks, such as obstacle avoidance, target tracking, and data analysis. The development of more efficient and powerful electric motors and propellers has increased the speed and agility of UAVs while reducing their noise and environmental impact. The development of sophisticated software for flight control, navigation, and mission planning has made it easier to operate UAVs and enabled them to perform a wide range of tasks.

The swarms of UAVs with intelligent monitoring mechanisms can reliably and quickly cover an intended area by using different parallel operating drones. Smaller UAVs can fly at a speed lower than 15 m/s. In contrast, large UAVs can fly at a speed of up to 100 m/s. Large UAVs can travel for hours, while smaller UAVs can fly for a limited time of 20–30 min. UAV's range refers to the distance from where it can be controlled remotely. The range differs from a few meters for small drones to hundreds of kilometers for larger drones [4].

8.1.4   UAV threats

Recently, the world witnessed a number of security incidents involving UAS, some of them related to nuclear facilities. "Drone Swarm" invaded Palo Verde Nuclear Power Plant in September 2020 twice. The drones flew in, through, and around the owner-controlled area, the security owner-controlled area, and the protected area. In France, several unidentified UAVs made flights in the restricted airspace over 13 out of 19 nuclear power plants in an apparently coordinated and organized manner, sometimes simultaneously over plants that are hundreds of miles apart [5] (Figure 8.1).

Unauthorized drones in the surroundings of aerodromes already represented a latent/potential risk for a couple of years, but it took the events at London Gatwick Airport in December 2018 to bring it to the attention of the public and the authorities. Between 19 and 21 December a total of 115 drone sightings over the airport of London-Gatwick were reported leading to the closure of its single runway. During the disruption, which lasted 33 h, over 1,000 flights had to be cancelled, thereby affecting some 140,000 passengers. On 3 February 2020, three out of four runways at Madrid Barajas airport were temporarily inoperable on a Monday morning, following a drone sighting, with 26 flights being re-routed. At Frankfurt airport runway operations and some flights were suspended twice within one month (8 February and 2 March 2020) due to the reported presence of drones [8].

There was a case of UAS's attack on an oil company in Saudi Arabia, in 2019, as well as a drone attack on petroleum tankers here in Abu Dhabi in 2022 (Figure 8.2).

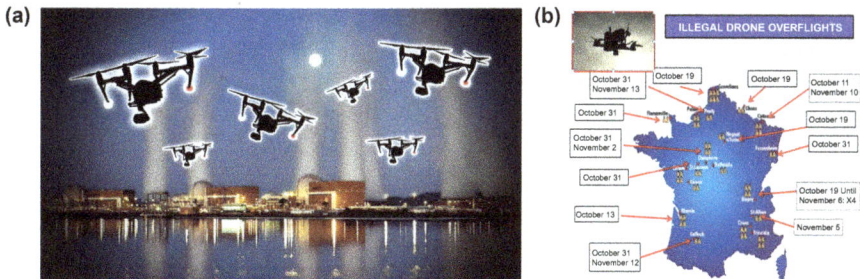

Figure 8.1   The examples of UAS's threat in nuclear power plants. (a) "Drone Swarm" invaded Palo Verde Nuclear Power Plant [6]. (b) Documented illegal drone overflights in France [7].

Figure 8.2   The case of UAS's attack on an oil company in Saudi Arabia [9,10]

8.2   C-UAV technology and EM characteristics

Many nations have initiated projects to equip some airports (civilian and military) with a C-UAS capacity, including UAS detection and/or neutralization capabilities.

8.2.1   Definition of C-UAS

In ED-286 which specifies minimum performance and interoperability requirements of a C-UAS System and was published by the European Organization for Civil Aviation Electronics (EUROCAE) in March 2021, anti-drone system is used as the term for C-UAS and is defined as follows: C-UAS refers to a set of technological and operational tools to monitor, detect, identify, record and enable response to unauthorized UAS activities; C-UAS may also include countermeasures capable to neutralize, or limit, potential risks, in a risk-based, balanced manner.

As a system of systems, a C-UAS system is typically composed of several complementary components, namely command and control, detection sensors and countermeasures, interfaces, and protocols to support interaction with other systems and stakeholders involved in the C-UAS stream and tools for information recording and post-analysis [11].

8.2.2   Configuration of C-UAS system

A C-UAS processing chain (Figure 8.3) is a framework for approaching the potential threat posed by UAS that can be used by technology developers and public safety officials alike [3].

A detection is a declaration that a UAS is in the presence of a sensor. Some systems, depending on how thresholds are configured, may report any object in its view as a detection (i.e., birds, commercial planes, etc.), or they may attempt to only alert the operator of objects deemed to be considered UAS, based on system capabilities and configuration. A location is a static estimated report or display of where a ground control station (GCS) or UAV is located at a given moment. The display to the operator of the C-UAS technology can take on many forms, e.g., a heat map display, quadrant alert, or circle to indicate estimated center and location

Figure 8.3   C-UAS processing chain [3]

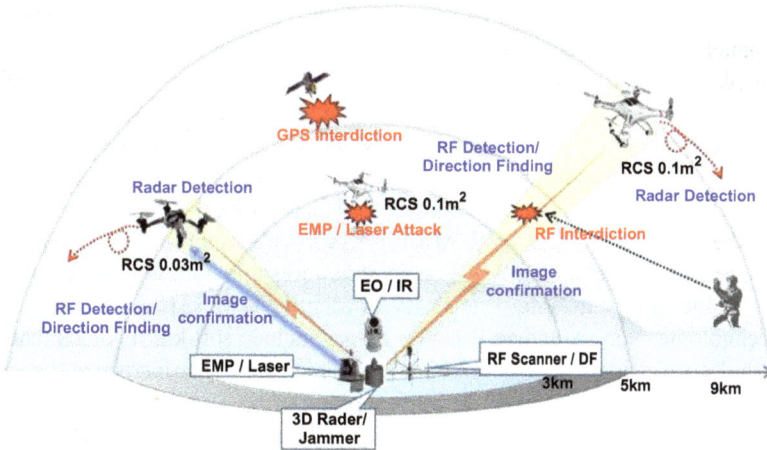

Figure 8.4   Anti-drone system (so-called counter-UAS system) [12]

error or line of bearing (LOB). A track is a compilation of location reports over a period of time. Tracks can be displayed for GCS and/or UAVs. Generally, it is displayed as a line or a sequence of dots. Classification is the assignment by the C-UAS technology (either autonomously or by an operator) of a potential target UAS to a high-level category such as UAS type, group, manufacturer, and/or specific communication protocol. Identification is the assignment by the C-UAS technology (either autonomously or by an operator) of a UAS to a more specific name or category, such as the physical address of its modem, or the exact make/model of the UAS. Mitigate is often used interchangeably with negate, interdict or neutralize. It describes the methods used to remove or reduce the threat posed by a UAS. These methods include technical means, such as RF or GPS jamming, spoofing/hijacking, and kinetic attacks [3].

Figure 8.4 summarizes and expresses the effective anti-drone system construction plan as a result of conducting interviews with experts in drone-related fields based on literature research [12].

8.2.3   Technology of C-UAS system

There are four modalities (or types) of sensors that are commonly used in C-UAS operations to detect, locate/track, and classify/identify UAS: Radar, Passive RF (sometimes referred to as electronic surveillance measures (ESM)), electro-optical (EO) and infrared (IR) cameras, Acoustic. Radars operate by transmitting a radio signal of known frequency and power in a focused direction and then detecting the reflected signal that is bounced back from the target. Two-dimensional (2D) radars provide direction and distance to the target, while three-dimensional (3D) radars also provide the target's altitude. Two-dimensional radars typically use a single antenna that rotates to cover the desired field of view. Three-dimensional radars use phased array stationary antenna panels with multiple internal antennas (array) on a single panel. Three-dimensional radars provide the target's elevation angle in the vertical plane in addition to its range and azimuth. Passive RF sensors rely on antennas to receive, and computers to analyze, RF signals associated with communications between the GCS and the UAV. Passive RF sensors analyze the radio signatures and modulations specific to UAS signals and are capable of identifying certain UAS models and manufacturers as well as locating the signal's transmission origin—the UAV and/or the GCS. Most C-UAS that rely primarily on passive RF sensors use libraries of known UAS radio signatures and compare detected signals to those in the library in order to classify or identify UAS [3].

Technologies that neutralize detected drones include soft kill methods that interfere with drones' missions with technologies such as radio interference and spoofing, and hard kill methods that destroy drone aircraft using lasers, high-power electromagnetic waves, and anti-aircrafts. Interference (Jamming) with radio waves disrupts the controller's radio operation or the drone's GNSS receiver, preventing the drone from moving to the desired goal. Radio jamming equipment has a limited range, so it must be close to the invading drone, and visible lines must be secured. Spoofing is a method of detecting the radio frequency of the drone by manipulating the communication protocol and transmitting stronger radio waves to seize control and forcibly land it. It intercepts the communication data link to control the target drone or injects a fake signal or malicious code into the GNSS receiver to disrupt location identification and force it to land. It is a technology that can emit high-power powerful electromagnetic waves to drones to burn electronic circuits or stop the operation of electronic components to crash drones, which can be a long-term and fundamental countermeasure. High-power electromagnetic waves can also affect other aircraft [12].

8.2.4 *Electromagnetic characteristics of jammers*

8.2.4.1 Investigation of characteristics of jammers

The operating frequency and RF output characteristics of jammers developed and commercialized were summarized as shown in Table 8.1.

In Table 8.1, the characteristics of radio link cutoff frequencies, signal modes, and types of major interference devices developed and commercialized to date can be summarized as follows:

• The frequencies used for jamming are some frequencies (433 MHz, 915 MHz, 2.4 GHz, 5.8 GHz) in the ISM band for drone-controlled communication and frequencies used for jamming GNSS (GPS, GLONASS) for positioning, typically ranging from 1.2 GHz to 1.6 GHz.

Table 8.1   *RF characteristics of Jammers*

• Operation	• ISM: 433 MHz, 915 MHz, 2.4 GHz, 5.8 GHz
• Frequency for jamming	• GNSS: GPS (L1, L2), GLONAS (L1, L2)
• Operation RF mode	• Others: L-Band (1–2 GHz), S-Band (2–4 GHz),
• Output	C-Band (4–8 GHz)
• 100 mW to 1,300 W	• Continuous wave, Frequency sweep, Pulsed up to 1,300 W
• Effectiveness distance	• Max. 10 km
• Max. 10 km	• Broadband, Multi-array
• Antenna type	• Directional/omni-directional/Directional
• Broadband (Log-periodic antenna)	
• Direction	

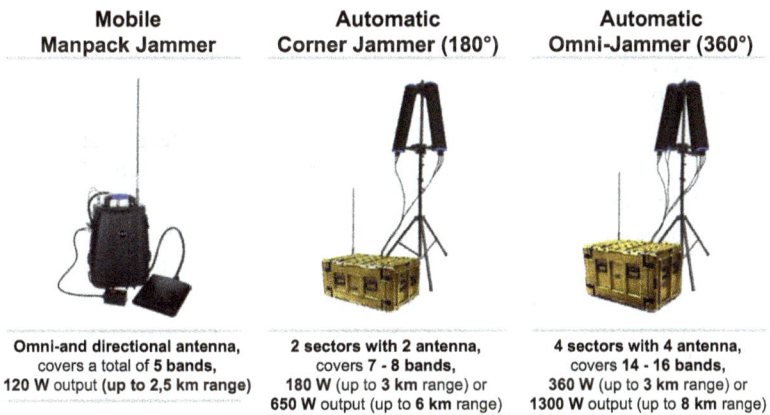

Mobile Manpack Jammer
Omni-and directional antenna, covers a total of **5 bands**, **120 W** output **(up to 2,5 km range)**

Automatic Corner Jammer (180°)
2 sectors with 2 antenna, covers **7 - 8 bands**, **180 W** (up to **3 km** range) or **650 W** output (up to **6 km** range)

Automatic Omni-Jammer (360°)
4 sectors with 4 antenna, covers **14 - 16 bands**, **360 W** (up to **3 km** range) or **1300 W** output (up to **8 km** range)

Figure 8.5   An example of the technical specification of jammers [13]

- Interference frequency bands mainly use continuous sinusoidal waves (CWs) of a specific frequency, broadband signals combining multiple frequencies, and signals that continuously vary frequencies at a constant magnitude within the interference frequency band.
- The types of obstruction devices can be divided into fixed types installed and operated in a specific place and portable types such as mobile or firearms (guns) installed in vehicles that can be carried and operated by humans.
- The effective distance was investigated from 500 m to 10 km.
- Company A's technical specifications for the jammer are shown in Figure 8.5 that it has a maximum output of 1,300 W, but this means the sum of the outputs of several frequency bands. For Company A, it can consist of up to 16 frequency bands and up to 8 antennas, and the technical specifications are as follows: Directional antenna: 2.45–2.50 GHz 100 W, 1.57–1.62 GHz (GPS L1

+GLONASS L1 Max 100 W, 5.70–5.90 GHz 25 W, Omnidirectional antenna: 433 MHz/100 W, 860–930 MHz 100 W, Depending on the number of antennas, the total output is 1,300 W. Most Jammers have an output of 120 W or less.

The electric field at a certain distance from a jammer can be calculated using the following Equation (8.1):

$$E = \sqrt{\frac{\eta_0 G P_t}{4\pi d^2}} = \frac{\sqrt{30 G P_t}}{d} \ (\text{V/m}) \tag{8.1}$$

where,

P_t: the transmitting radiated power,
G: the transmitting antenna gain,
d: the distance between the transmitting antenna and receiving antenna,
η_0: the characteristic wave impedance in free space.

8.2.5 Increasing aspects of HPEM threats to the future society

The RF active equipment/device of the C-UAS system can affect, degrade, or block performance on the other communications, navigation, and surveillance (CNS) systems, and infrastructure. If a high-power electromagnetic pulse is used as a type of active RF device, a powerful HPEM wave is sent in the direction of the UAV. This can disable some of the electronic components, causing the UAV to land. The main disadvantage of this technique is that this pulse could damage some of the surrounding electronic infrastructure. Similar to the "nondestructive" jamming technique, this approach would be dangerous for security sensors and equipment.

The fact that HPEM sources are not only used on the battlefield as military weapons but also in public may pose a threat to current and future societies where information and telecommunication are advanced, even if their place of use is limited and strictly regulated by law.

First, it is possible to consider the situation of mission drones used for public safety. The need to use drones in these areas is growing more than ever. Many first responders, including police officers, firefighters, search and rescue teams, and other public safety organizations, use various drones and payloads in their daily tasks to perform their duties better, safer, and faster. It is critical that these drones for public safety missions protect their wireless communication capabilities from external high-level EM threats in performing a safe and complete mission.

In addition, consumer and commercial robots typically operate in public places, commercial environments, and homes near humans, so it is important to increase the safety of robots operating in human-friendly environments. These robots typically use Wi-Fi or 5G communication as a wireless communication feature. These wireless communication functions are feared to be vulnerable to external HPEM threats.

As the number of applications using high-power electromagnetic waves for special purposes increases, the threat of external electromagnetic waves to wireless communication functions such as drones and robots for public safety missions is also increasing. Therefore, it is necessary to evaluate and respond to how the jammers currently used in the C-UAS system or HPEM sources available in the future can affect the surrounding equipment or systems.

8.3   International standardization for vulnerability analysis against EM threats

Methods for assessing whether equipment or systems are affected by external EMs are covered by some IEC technical committees. However, there are few cases in which specific evaluations have been made in the proposed method. Here, each progress is reviewed and advanced methods are sought to appropriately evaluate the impact of equipment or systems by high-power jammers or other HPEM sources applied to C-UAS systems in the future.

8.3.1   IEC TC77

First, the international standard IEC 61000-1-2 has been prepared by Technical Committee 77 on IEC Electromagnetic compatibility. It established a methodology for the achievement of functional safety only with regard to electromagnetic phenomena. This methodology includes the implication it has on equipment used in such systems and installations. This describes hazard and risk analysis considering electromagnetic environment assessment. In this standard, assessing the EM environment is to determine the most severe electromagnetic environment by either designers, manufacturers, installers or users of the system; to choose only equipment for use in an electromagnetic environment equal to or more severe than the maximum environment. In this procedure, equipment manufacturers typically specify that their equipment has been tested to applicable EMC standards and comply with them at specified levels. If the known application environment exceeds the equipment specifications, appropriate means shall be applied to ensure adequate performance. Therefore, we can understand that the EM environment assessment is to compare both levels between the maximum environment and specified level of applied EMC standards [14].

Second, IEC/TS 61000-5-9, which is a technical specification, has been prepared in 2009 by subcommittee 77C. The technical specification presents a methodology to assess the impact of high-altitude electromagnetic pulse (HEMP) and HPEM environments on electronic systems (see Figure 8.6). This work is closely related to the evaluation of EMC system-level susceptibility. Typical systems have external connections, wired or wireless, and the assessment of these is included within this specification [15].

Section 6.2 of the standard addresses the characterization phase of subsystems and equipment. The functional and topological descriptions provided are crucial for understanding the propagation of front-door and back-door coupling within the

Subsystems &
Equipment
Characterization
Phase

Identify Functionally Critical
Subsystems and Equipment

Susceptibility or
Immunity Data
Available?

Y

Characterize Data as
Waveform Norms

N

Estimate EUT immunity/
Susceptibility

System Analysis
Phase

Perform System
Susceptibilty Analysis

Select critical system interface and
coupled stress measurement points

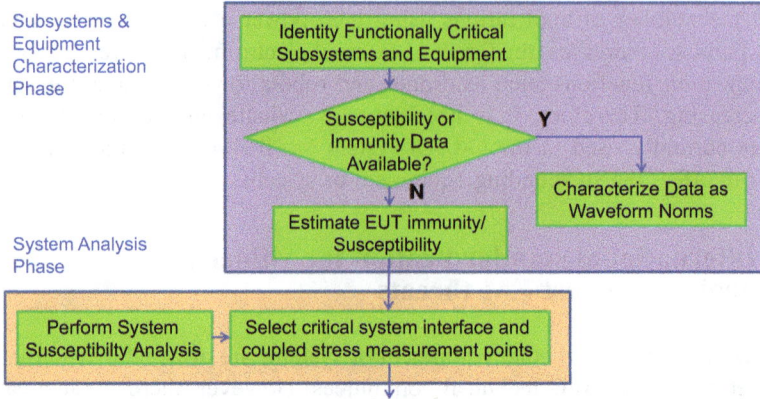

Figure 8.6   Assessment methodology flow chart in IEC 61000-5-9

system, which is key to developing effective hardening strategies. These descriptions also aid in determining whether HEMP or HPEM-induced effects should be classified as immunity or susceptibility during the susceptibility assessment phase. During this phase, it is important to gather all relevant information regarding immunity or susceptibility, ideally from electromagnetic compatibility (EMC) test data. However, obtaining HEMP-specific data from standard EMC tests often presents practical challenges [15].

8.3.2   ITU-T

ITU-T K.81 is a recommendation for a high-power electromagnetic immunity guide for telecommunication systems. This recommendation presents guidance on establishing the threat level, the physical measures, and establishing the vulnerability of equipment. In this recommendation, we can see the threats and their peak levels in electric fields at certain distances. It also describes the immunity level of typical IT devices according to the type of EM emanation. Then, Equation (8.2) for the determination of EM mitigation levels is as follows. EM mitigation level can be replaced by the effects of EM threats [16].

$$(\text{EM mitigation level}) = (\text{Threat level}) - (\text{Vulnerability level}) \tag{8.2}$$

8.3.3   Applied to C-UAS system

Back to the C-UAS system again, EM effects due to radar or an RF jammer can be calculated by this equation.

$$\text{Effects of EM} = (\text{Electric fields}) - (\text{Vulnerability level}), \text{in dB} \tag{8.3}$$

In (8.3), the electric field can be measured at a distance or simulated, peak or average, and vulnerability level typically can be obtained from EMC data of equipment. However, it needs to re-establish the level of electric field from the source for cost-effective protection measures giving a weighting factor for the peak electric field depending on the type of signals.

Military Handbook 235-1C (2020), Military Operational Electromagnetic Environment Profiles provide information for use in tailoring and supplementing the electromagnetic environment (EME) levels specified in Mil-Std-464C and the radiated susceptibility requirement RS103 of Mil-Std-461G. It provides the relationship between average and peak power calculation [17].

Pulse-modulated signals, typically from radars, have differences between peak and average power. The average power is determined by the ratio of time-on to time-off over an interval, which is called a duty cycle.

$$P_a = P_p \times d.c. \tag{8.4}$$

$$d.c. = \frac{pw}{pri} \text{ or } d.c. = pw \times prf \tag{8.5}$$

where,

$d.c.$: the duty cycle,
pw: the pulse width (seconds),
pri: the pulse repetition rate interval (seconds),
prf: the pulse repetition rate frequency (Hz),
P_a: the average power,
P_p: the peak power.

In (8.4) and (8.5), for comparison, we can convert peak power to average power by applying the duty cycle factor of the pulse. It can be one method that the power equation can be replaced by the electric field equation.

$$E_a = E_p \times d.c. \tag{8.6}$$

$$\text{Effects of EM} = (E_p \times d.c.) - (\text{Vulnerability level}), \text{in dB} \tag{8.7}$$

So, when we assess the EM effects of equipment or systems from HEMP or HPEM according to the relevant standards, we need to apply a weighting factor like duty cycle, considering the characteristic of a waveform, for comparison.

8.4   Conclusion

There has been a significant growth in the legitimate use of Unmanned Aerial Systems over recent years. This is anticipated to continue as new and innovative uses are found and the capabilities of UAS continue to develop. However, there is also a potential for drones to be used carelessly or for malicious purposes, including terrorist and other criminal acts. As concerns grow around the potential security threats drones may pose to both civilian and military entities, a new market for counter-UAS technology is rapidly emerging. Several security incidents involving UAVs have been witnessed around the world, making them a real threat. Many countries have begun implementing C-UAS systems. Most C-UAS systems use radar and RF jammers for detecting and neutralizing UAS. However, due to their relatively high RF output, they can affect existing critical systems, subsystems, or peripheral equipment.

This chapter examines the technology trends of UAS and C-UAS systems and explores the electromagnetic characteristics of C-UAS systems. It measured the electric field from jammers of C-UAS systems to investigate how high levels could affect other equipment or systems around, especially front-door coupled. The chapter also looked at IEC standards and other standards for assessing EM vulnerability in national critical infrastructures caused by EM or HPEM sources in terms of HPEM threats to mission drones and robots. When one assesses the EM effects of equipment or systems from HEMP or HPEM according to the relevant standards, one may apply a weighting factor like duty cycle, considering the characteristic of a waveform, for comparison. It needs further study. It would be good for the IEC's corresponding technical committee to deal with this issue.

References

[1] David Hodgkinson and Rebecca Johnston, Guiding principles for drones: A starting point for international regulation, *Perth International Law Journal*, 3, pp. 158–184, (2018).
[2] Office of the Secretary of Defense, Unmanned Aircraft Systems Roadmap 2005–2035, page 14, Department of Defense, 2005.
[3] Counter-Unmanned Aircraft Systems – Technology Guide, September 2019, the U.S. Department of Homeland, Security, Science and Technology Directorate, National Urban Security Technology Laboratory (NUSTL).
[4] Hazim Shakhatreh, Ahmad H. Sawalmeh, Ahmad H. Sawalmeh, *et al.* Unmanned aerial vehicles (UAVs): a survey on civil applications and key research challenges, *IEEE Access*, 7, pp. 48572–48634, (2019).
[5] Alexander Solodov, Adam Williams, Sara Al Hanaei, and Braden Goddard, Analyzing the threat of unmanned aerial vehicles (UAV) to nuclear facilities; *Security Journal*, 30(1), pp. 305–324 (2018).
[6] https://www.thedrive.com/the-war-zone/34800/the-night-a-drone-swarm-descended-on-palo-verde-nuclear-power-plant.
[7] André Ranson, The 2014 UAV threat to French nuclear power plants, *National Security and the Future*, 18(1–2), pp. 125–142 (2017).
[8] EASA, Drone Incident Management at Aerodromes Manual, 8 March 2021.
[9] https://www.scmp.com/business/companies/article/3027525/impossible-proceed-aramco-ipo-investors-fear-ongoing-attacks.
[10] https://www.abc.net.au/news/2019-09-18/us-believes-attack-on-saudi-arabia-came-from-southwest-iran/11522678.
[11] EUROCAE, ED-286, Operational Services and Environment Definition for Counter-UAS in Controlled Airspace, 2021.

[12]   Soon-Phil Hwang and Doo-hwan Kim, A study on the establishment of anti-drone system for the protection of national important facilities, (Korean) *Journal of Digital Convergence*, 18(11), pp. 247–257 (2020).

[13]   AARTOS-Sector Jammer Datasheet in Germany (www.aaronia.de).

[14]   IEC 61000-1-2 Ed 1.0: 2016, Electromagnetic compatibility (EMC) – Part 1-2: General – Methodology for the achievement of functional safety of electrical and electronic systems including equipment with regard to electromagnetic phenomena.

[15]   IEC/TS 61000-5-9 Ed 1.0: 2009, Electromagnetic compatibility (EMC) – Part 5-9: Installation and mitigation guidelines – System-level susceptibility assessments for HEMP and HPEM.

[16]   ITU-T K.81, High-power electromagnetic immunity guide for telecommunication systems, June 2016.

[17]   MIL-HDBK-235-1C (2010), Part 1C – Military Operational Electromagnetic Environment Profiles, US DoD, 2010.

Chapter 9

Real-time shielding compromise and detection

Eric Easton[1], Ryan Marietta[1] and Richard Hoad[2]

As with many security-centric initiatives, a framework can aid in evaluation and management decisions. CenterPoint Energy chose to adopt the NIST framework from cyber security to evaluate high-power electromagnetic (HPEM) mitigation requirements. The NIST framework utilizes five functions, including identify, protect, detect, respond, and recover. This chapter will focus mostly on the detect and respond aspects of the framework. A successful response to an HPEM event will require the ability to detect and capture data for analysis. Second, the response will be determined by the effectiveness of mitigations and resulting impacts on the system. Therefore, ensuring the mitigations are meeting the design basis over the life of the system is critical. The effectiveness of HPEM mitigation depends on a functional mitigation system beyond the initial installation; however, testing on an electric delivery system comprised of hundreds of substations could result in pro-hibitively high maintenance costs. The successful detection and monitoring of shielding effectiveness for hundreds of locations led CenterPoint Energy to colla-borate with QinetiQ to solve both design and operational challenges.

9.1 Detection of HPEM threats and shielding compromises

9.1.1 Introduction

This sub-section introduces the Totem® threat detection and prototype Totem Shielding Compromise (SC) solution. The design choices made in the development of the prototype Totem SC solution are discussed. Totem SC is still in development and has only undergone some basic performance evaluation to date. This perfor-mance evaluation has, due to the coronavirus pandemic, had to be undertaken on a setup that is not fully representative of the intended application setup. The intended application is for the monitoring of the CenterPoint Energy Resilient Digital Substation Electromagnetic Pulse (EMP) module introduced earlier. The Resilient Digital Substation EMP module is a high-quality shielded enclosure or cabinet with

[1]CenterPoint Energy, Houston, TX, USA
[2]QinetiQ Ltd., Cody Technology Park, Farnborough, Hants, UK

approximate dimensions of 2 m high by 1.5 m wide by 1.5 m deep. The compact design was accomplished by replacing the traditional metallic control cables with fiber optic cables for protection and control [1].

Efforts related to HPEM mitigation began with the CenterPoint Energy Control Center before progressing to mitigation designs for substation infrastructure. After exploring more traditional designs for fixed installations, which would have resulted in a shielded area of approximately 3.5 m high by 6 m wide by 9 m deep; the Resilient Digital Substation EMP module was selected as the most efficient mitigation approach for electric substations. The smaller shielded volume is also easier to test and maintain while allowing for installation in existing substation control buildings shown in Figure 9.1 [1].

A resilient response to an HPEM event will require both mitigation and detection. Detection provides situational awareness and aids in determining the type of response plan to initiate. Although hardened infrastructures are expected to remain operational, effects on unprotected public systems would still create the need for operational adaptation. Therefore, a system that provides awareness of an HPEM event will serve as the first communication to the control center to begin deploying field resources. This notification will only be possible if the monitoring device is capable of withstanding the HPEM event without upset or damage.

Figure 9.1   Resilient digital substation EMP module installed in substation

The design basis required consideration of both high-altitude electromagnetic pulse (HEMP) and intentional electromagnetic interference (IEMI); therefore, the appropriate frequency ranges were examined for both detection and system survivability. Situational awareness is also needed with respect to the shielding effectiveness of installed systems. Ongoing testing ensures that installed mitigation efforts have not experienced shielding degradation which would undermine response planning in the event of a system-wide event. Decisions related to staffing levels, inspection protocols, and safety stocks all depend on the aggregated shielding effectiveness of disparate shielding systems at multiple substations and control centers. The ability to review real-time shielding effectiveness data across the entire system can aid in prioritizing resources for shielding repairs and allocation of on-hand materials to the most critical sites. In this manner, real-time shielding effectiveness ensures optimized systemic resiliency against HPEM events.

Should an event occur technicians would be dispatched to the appropriate substation locations. Given the level of attenuation incorporated into the Resilient Digital Substation module, units without shielding degradation would not suffer damage from events conforming to the unclassified waveforms based on third party empirical testing of the Resilient Digital Substation EMP module. Should a unit be suffering from compromised shielding at the time of an HPEM event the data collected by the detector will provide a significant benefit in the inspection process. One of the challenges with HPEM threats can be a lack of visible evidence of the system failure mode. The challenge in determining a root cause may delay the initiation of the restoration effort and exhaust available resources if the number of locations requiring investigation is excessive. In some scenarios, the power system operator could be unaware of the damage until the substation equipment is required to operate in response to a fault in the power system. The Totem SC detection system will provide information that can be compared to data collected while completing failure mode testing of each component comprising the Resilient Digital Substation module. A lookup table based on the environmental conditions during the HPEM event will provide an efficient means to determine which components may have been affected and what type of failure modes should be investigated greatly reducing the time to return systems to service. Should units with compromised shield exist, they could be prioritized greatly reducing the number of technicians required and allowing for these individuals to be reallocated to other emergency response tasks.

Optimizing costs is a common concern for utility companies as they pursue resiliency and readiness for high impact low frequency (HILF) events. Many large electric system operators have responsibility for hundreds of substations which would result in near-daily shielding effectiveness testing of any one site per day assuming a typical annual testing schedule per site. For example, CenterPoint Energy oversees more than 300 substations and following a system-wide hardening effort would have to test more than one site per day to ensure each site was tested on an annual basis. Without the deployment of the Totem SC solution, maintenance would require an onsite technician test per Resilient Digital Substation EMP module for one day per technician. Testing would include an IEEE 299 [2] test to determine if the shielding effectiveness was meeting the design basis. To limit emissions within the in-service substations this test is conducted with the transmitter inside the shielded enclosure and the receiver 1 m from the enclosure. Measurements are then taken on each side of the enclosure to determine the need for any adjustments to the HPEM mitigation components such as doors, wave guides below cutoff, etc. This estimate considers travel to the site, test equipment setup, calibration, testing, demobilization and travel back to the work center. When compared to online shielding compromise systems, annual testing plans will be significantly more costly.

The use case for an online shielding compromise and detection system was based on the aforementioned benefits associated with situational awareness, maintenance cost reduction, and post-HPEM event emergency operations. The specific design requirements and adaptation to a compact shielded enclosure led to the development of the Totem Shielding Compromise (SC) solution. After establishing the use case and defining a design basis, CenterPoint Energy began collaborating with QinetiQ in their development of the Totem SC system.

9.1.2   HPEM threat detection

A HPEM threat detection capability has been in development at QinetiQ for some time and this has culminated in the production of the Totem detector system. The Totem detector provides prompt detection of radiated transient E-fields that could pose a threat to the function of electronic systems [3]. The detector has a novel broadband spiral antenna and uses a logarithmic receiver [4]. The detector has onboard flash memory for storing event data, an integral uninterruptable power supply, fiber-optic connections for event data transmission, and can be configured to interface with a 24/7 web service. The Totem detector and the web-based reporting user interface are shown in Figure 9.2.

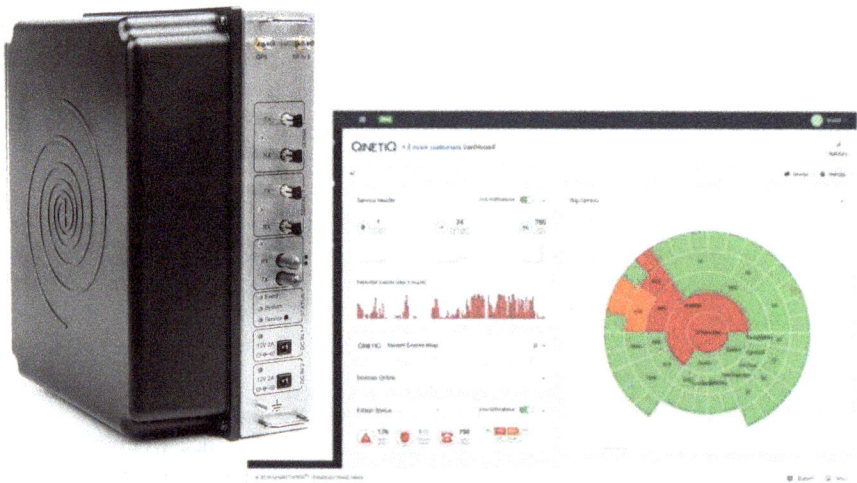

Figure 9.2   Totem® Threat Detection and reporting

Detected Events
All

1M All   📅 Oct 01,2020–Nov 05,2020

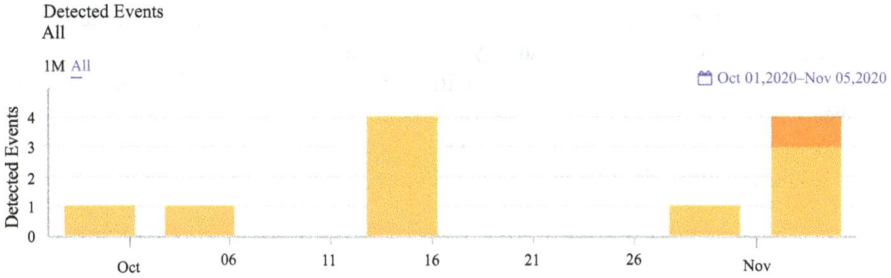

*Figure 9.3   Totem detected HPEM event data from a Scandinavian HV
sub-station*

The characteristics of this detection system are as follows:

- Frequency Range: 10 MHz to 10 GHz
- Instantaneous bandwidth: $\sim$100 MHz—proven to detect Hyperband environments such as transient waveforms with $\sim$100 ps rise time and $\sim$200 ps pulse widths

The data gathered by Totem can be used to identify the prevalence of radiated EM threats and guide the adoption of cost-effective protection. The installation of detectors also serves to raise awareness of the plausible existence of the threat to system operators. Figure 9.3 shows data gathered from the deployment of the Totem detector at a high voltage (HV) electrical substation site in Scandinavia.

The data shows that the Totem threat detection system was able to detect and report HPEM events occurring during a 6-month trial at the substation site. In this case, the detected events were correlated with non-malicious activity, specifically, the events were correlated with the opening of an HV circuit breaker on the site.

9.1.3   The need for shielding compromise detection

The use of shielding to protect sensitive electronic systems from HPEM environments is well known. The shielding provides attenuation, which is articulated in terms of the dB reduction in the magnitude of an EM field outside a shielded volume to the magnitude inside the shielded or protected volume.

The effectiveness of a shielded enclosure depends on many parameters. In theory, a shielded enclosure may be designed to produce attenuation ranging from a few dB to over 100 dB in a frequency range of up to 10 GHz and beyond. However, in practice, the effectiveness of an enclosure with apertures and penetrations of all types will be reduced and limited by these apertures and penetrations [5].

The performance or effectiveness of the shield is essential to the protection and resilience of sensitive electronics. However, the measurement of shielding effectiveness of a shielded enclosure can be technically challenging and time-consuming. Measuring shielding effectiveness as described in standards (e.g., [2,6]) typically requires the use of a radiated test system and, depending on the volume of the shielded enclosure can take several days to carry out.

The shielding effectiveness or shielding performance is defined as the ratio between the field strength (Electric or Magnetic) at a given distance from the source without the shield interposed and the field strength with the shield interposed. The equation for the shielding effectiveness (S_E) for electric fields is:

$$S_E = -20\log_{10}\left(\frac{E_{inside}}{E_{incident}}\right) \ (dB)$$

When the shielded enclosure is used to protect an operational facility that is part of critical national infrastructure it can be difficult to arrange time for shielding effectiveness testing to be carried out on a regular basis and it may even become impractical to carry out a measurement as the space within the facility is filled with operationally critical equipment.

Typically, a shielding effectiveness test is carried out when the shielded enclosure is commissioned and thereafter only verified by measurement once per year at most. This means that if the shield has been violated or compromised in some way there could be a significant period of time where the shield is not providing the required protection.

Shielding violations and compromises

- Radiated violations — Slots/apertures created or caused by:
  - Degradation/corrosion of gaskets and finger stock used for doors and mating surfaces;
  - Incorrect application of gaskets (i.e. non-conducting environmental seals used in place of conductive shielding type gasket);
  - Poor compression of doors and other mating surfaces;
  - Doors or hatches left open or ajar.

- Conducted violations
  - Degradation/corrosion of cable shield bond;
  - Inadequate termination of cable shields to the shield surface (e.g., high impedance bond; pigtail);
  - Installation of non-shielded conductive penetration (e.g., wire, cable, pipe, or bolt);
  - Failure of filters or non-linear transient protection.

One way of handling this issue of shield degradation is to include a "margin" in the attenuation performance specification. Shielding effectiveness of radiated Electric fields above 100 MHz can be specified at 80 to 100 dB for high altitude EMP (HEMP) E1 waveform, for instance [7]. However, a simple analysis can show that ∼30 dB is sufficient to provide radiated protection for modern electronic systems from HEMP E1, therefore this specification includes a significant margin, perhaps as much as 50 to 70 dB.

While the inclusion of a protection margin is tried and tested it does not guarantee that shielding violations and compromises will not detrimentally affect the protection and importantly the degradation may only be detected at the next measurement survey date.

The semi-continuous, scheduled, commanded, and automated monitoring of shielding performance is therefore an important need and QinetiQ was tasked to provide a solution as an add-on to the Totem threat detection solution.

9.1.4 The Prototype Totem Shielding Compromise detector

The Totem threat detector was already equipped with most of the elements necessary to provide a shielding compromise detection service such as the wide-band receiver and the data communications and reporting sub-system. CenterPoint Energy had already procured the threat detector functionality. The need therefore was to develop a way of detecting shielding compromises using as much of the Totem threat detection capability as possible.

9.1.4.1 Design choices

Certain design choices were necessary to implement the Totem SC system and a selection of these choices and the design decisions made are summarized below:

- Radiated fields or Conducted injection.
- Reference signal magnitude and frequency range—to perform an A-B measurement as required to ascertain the shielding attenuation, a Reference signal must generated.
- Reference signal generator location—inside or outside of the EMP module shielded enclosure.

The first design choice was to ascertain if a radiated field test was needed or if a conducted injection test would be adequate. The primary concerns with the radiated field test were: the potential need for an RF transmission license, the potential for the radiated field to interfere with equipment within or outside of the shielded enclosure, and the stand-off distance required between the transmitting and receiving antennas and the enclosure wall. The concerns associated with the conducted injection method were: sensitivity, could we achieve the required amplitude dynamic range? Given the above, it was decided to investigate using a conducted/injection method. A photograph of our coupler injection device is shown in Figure 9.4.

Two identical injection/couplers per installation are needed one for the inside of the shielded enclosure and the outside of the shielded enclosure. The only modification required to the shielded enclosure is the fitting a high-quality coaxial bulkhead connector. The objective of the coupler is to induce an RF skin current in the surface of the enclosure and to measure the reciprocal skin current on the other side of the shield surface.

The frequency range for the reference signal generator was selected to sweep across the frequency range 75 to 145 MHz. This was a pragmatic decision based on several factors:

1. The major EM threat of concern was HEMP E1. Approx. 90% of HEMP E1 energy is below 100 MHz [8].

Figure 9.4   Totem SC coupler/injection device fitted to a shielded enclosure

2.   Practical availability of components.
3.   Slot apertures and uncontrolled cable violations will likely have resonance within this range [5].

The location of the reference signal generator was selected to be outside of the shielded enclosure. Through experimentation, it was found that there was no practical difference in performance but the reference signal generator added extra volume to the Totem system and space within the enclosure was at a premium.

A schematic diagram of the installation is shown in Figure 9.5.

9.1.5   The Totem Shielding Compromise detector performance

9.1.5.1   Amplitude dynamic range test

Figure 9.6 shows the amplitude dynamic range performance of the system when a direct connection is made between the injection and receiving devices using the reference signal sweep generator.

It can be seen that the amplitude dynamic range is of the order of 70 dB. The sweep time is settable and was set to 10 seconds.

9.1.5.2   Shielded enclosure (>80 dB) performance test

The performance of the prototype Totem SC detector was evaluated while fitted to a high-performance shielded enclosure (semi-anechoic chamber) located in QinetiQ Farnborough, UK. Figure 9.7 shows the detected signal when the shielded enclosure door is closed.

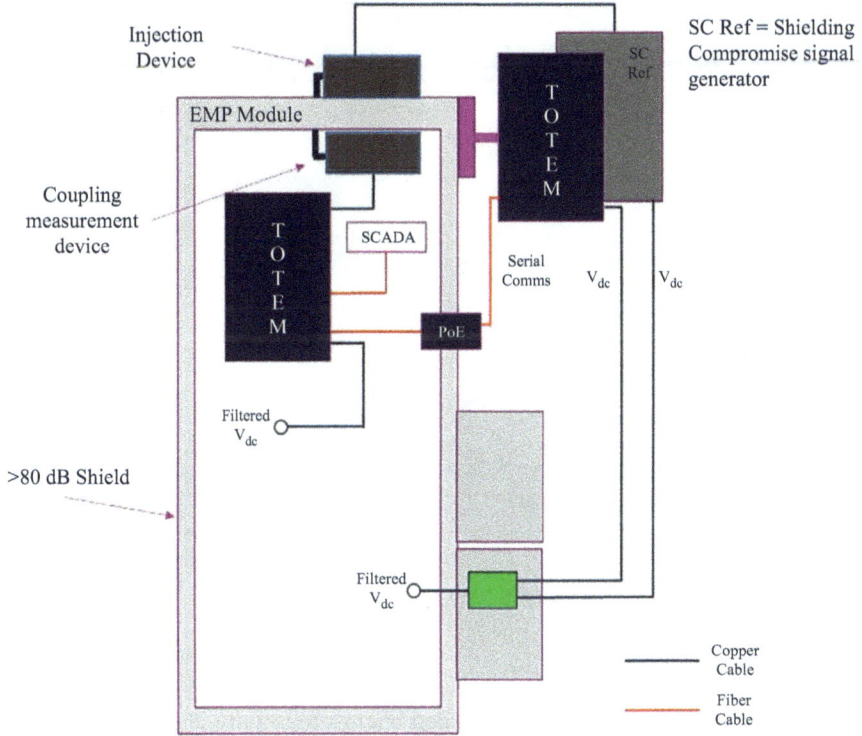

Figure 9.5 Schematic diagram of the Totem Shielding Compromise installation

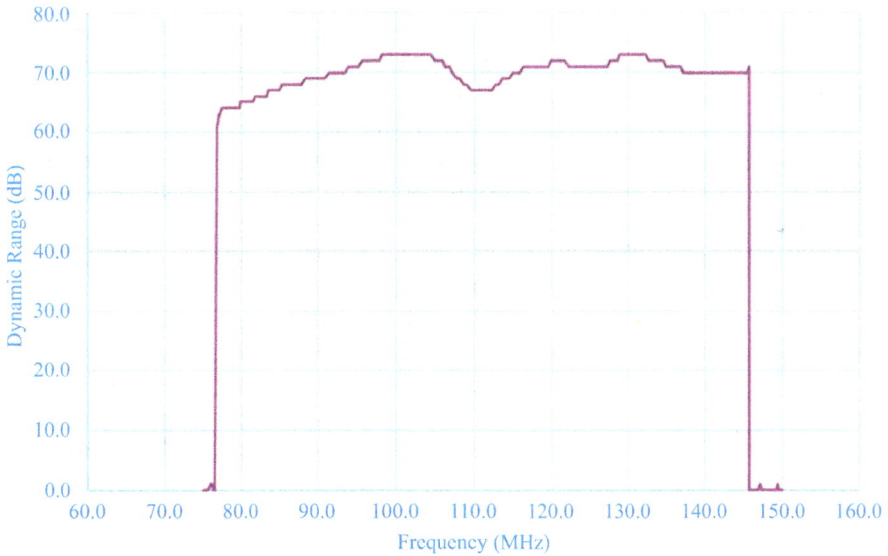

Figure 9.6 Amplitude dynamic range during a frequency sweep

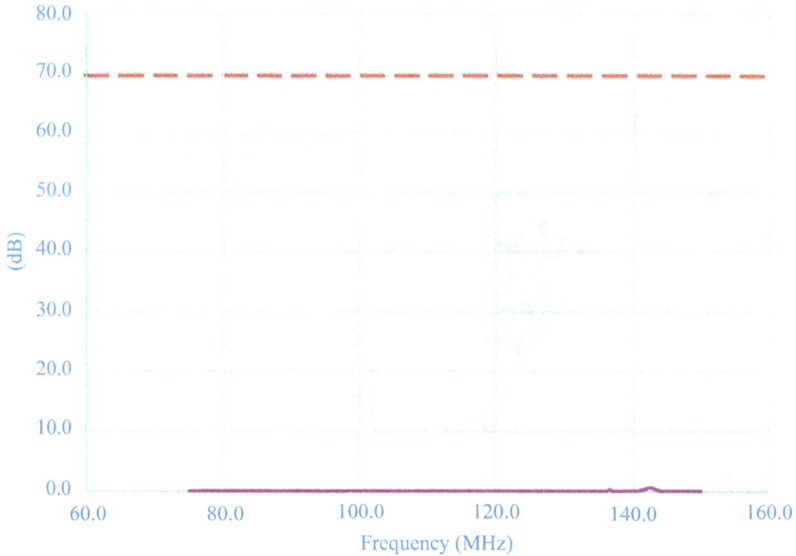

Figure 9.7   Shield closed test

The detected signal when the shielded enclosure is fully closed is barely distinguishable. This result is to be expected when the shield is performing. The red dotted line indicates the maximum possible amplitude dynamic range.

9.1.5.3   Slot aperture test

Figure 9.8 shows the detected signal for three different situations: (1) door firmly closed, (2) door compression released allowing a slot aperture to form and (3) the door ajar with the door opening less than 50 cm.

It was observed that releasing the closing mechanism thus removing the compression of the shielded enclosure door was sufficient to provide a detected signal of >30 dB. The detected signal increased to a maximum of 40 dB when the shielded enclosure door was opened by 50 cm. The "door open" case did not achieve a signal of the full 70 dB dynamic range of the reference signal. This was investigated and found to be due to the path length from the point of injection through the open door and back to the coupling measurement point. In this installation which is not topologically representative of the intended installation the path length was approximately 7.5 m as the enclosure is physically large and the penetration point is a considerable distance ∼3.5 m from the door. It is anticipated that this technique would naturally work better where path lengths are shorter, which is the case for the CenterPoint EMP module where the maximum path length could only be 4 m in the worst case. The suspected cause of the reduction of amplitude dynamic range is due to surface current propagation loss, likely dominated by resistive losses.

9.1.5.4 Conducted violation test

Figure 9.9 shows the performance of the Totem SC prototype with a simulated conducted violation introduced into the shielded enclosure. The cable violation was simulated by removing a bulkhead connector from the shielded enclosure

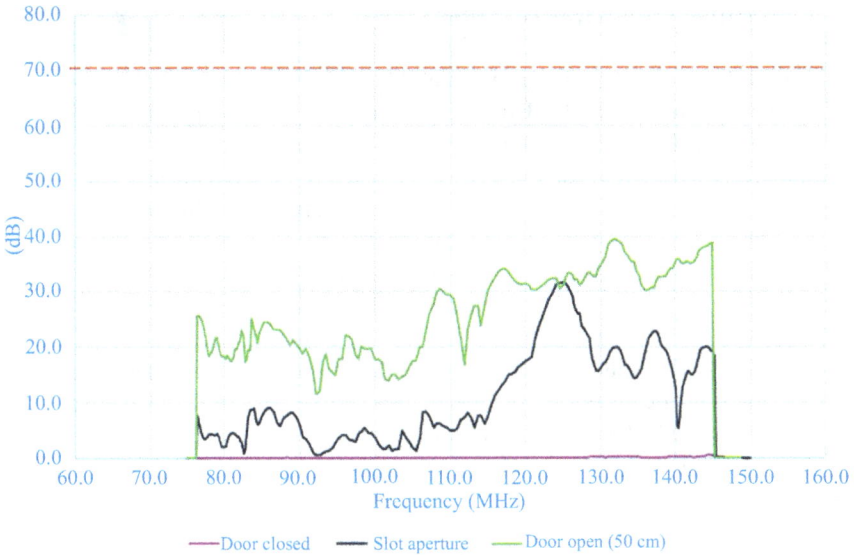

Figure 9.8   Slot aperture and door open detection test

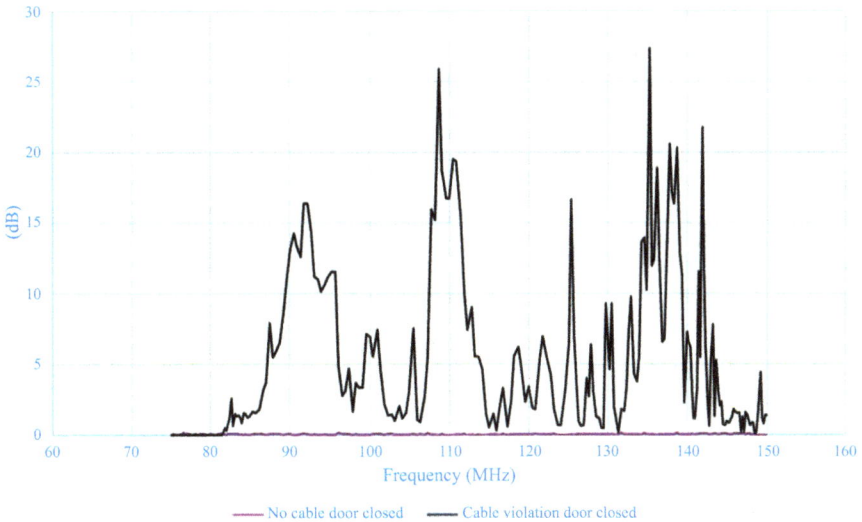

Figure 9.9   Conducted violation detection test. Again, a strong signal was detected. The path length for this test was ~4 m.

penetration panel and routing an Ethernet cable through the open hole so that at least 4 m of Ethernet cable was present within and outside of the chamber. The Ethernet cable was connected to equipment within the shielded enclosure to fully represent the likely condition.

9.2   Summary and limitations

The Totem threat detector is a new capability that is starting to be proven in real operational environments. We have developed a new prototype variant of Totem SC which is capable of providing detection of common shielding violations. It is important to note that the Totem SC is not a measurement device and a measure of the shielding effectiveness nor the absolute magnitude of any shielding degradation can be inferred from the performance test results. This is because the reference test used for Shielding Effectiveness within the standards is a radiated field test.

It is likely to be possible to calibrate the Totem SC solution against a radiated field test. A measurement of the transfer function of the radiated field to shield skin current would enable this calibration, though the performance of Totem SC and the transfer function are critically dependent on the dimensions and quality of the shielded enclosure Totem SC is fitted to.

An indication of how the detection data could be utilized is provided in Figure 9.10.

Note that degradation of non-linear transient protection devices which may be used in an HPEM-protected installation cannot be evaluated with the Totem SC system if the transient protection device fails to open the circuit. Failure and degradation of transient protection devices are best evaluated through the injection of a high voltage or current transient such as the pulse current injection (PCI) test.

Our next steps are to evaluate the performance of Totem SC installed on the intended CenterPoint Energy EMP module.

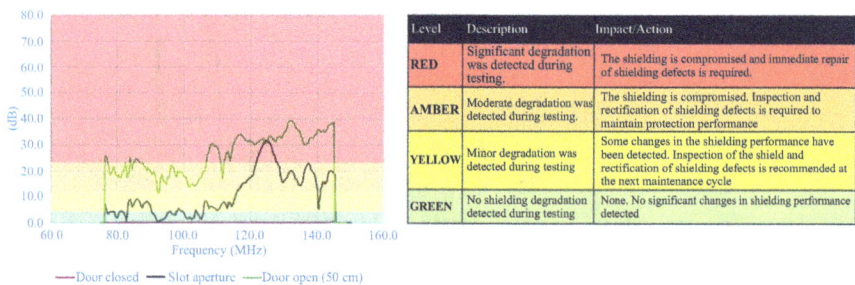

Level	Description	Impact/Action
RED	Significant degradation was detected during testing.	The shielding is compromised and immediate repair of shielding defects is required.
AMBER	Moderate degradation was detected during testing.	The shielding is compromised. Inspection and rectification of shielding defects is required to maintain protection performance
YELLOW	Minor degradation was detected during testing	Some changes in the shielding performance have been detected. Inspection of the shield and rectification of shielding defects is recommended at the next maintenance cycle
GREEN	No shielding degradation detected during testing	None. No significant changes in shielding performance detected

—Door closed — Slot aperture —Door open (50 cm)

Figure 9.10   A possible interpretation of the Totem SC data

References

[1] Easton E., Marietta R., and Hoad R., GLOBALEM 2022, Real-time Shielding Compromise and Detection, Abu Dhabi, UAE, 2022.

[2] IEEE Std 299.1, IEEE Standard Method for Measuring the Shielding Effectiveness of Enclosures and Boxes Having All Dimensions between 0.1 m and 2 m, 2013.

[3] Herke D., Chatt L., Petit B., and Hoad R., Lessons Learnt From IEMI Detector Deployments, EUROEM 2016, Imperial College, London, UK, July 2016.

[4] Hoad R., and Herke, D. L., Electromagnetic Interference Indicator and Related Method, *International Patent Publication Number*: WO17/125465 A1, filing date 19 January 2017.

[5] IEC 61000-5-6 Ed. 1.0 (2024-04-05): Electromagnetic Compatibility (EMC) – Part 5-6: Installation and Mitigation Guidelines – Mitigation of External EM Influences.

[6] IEC 61000-4-23 Ed. 2.0 (2016-10-20): Electromagnetic Compatibility (EMC) – Part 4-23: Testing and Measurement Techniques – Test Methods for Protective Devices for HEMP and Other Radiated Disturbances.

[7] Mil-Std-188-125-1, High-Altitude Electromagnetic Pulse (HEMP) Protection for Ground-Based C4I Facilities Performing Critical, Time-Urgent Missions, Part 1, Fixed Facilities, 2005.

[8] IEC 61000-2-9 Ed. 1.0 (1996-02-19): Electromagnetic Compatibility (EMC) – Part 2: Environment – Section 9: Description of HEMP Environment – Radiated Disturbance.

Chapter 10

Financial comparative analysis of substation HPEM mitigation designs

Eric Easton[1] and Ryan Marietta[1]

CenterPoint Energy completed the design of an effective, cost-efficient solution for high-power electromagnetic (HPEM) mitigation to be used in electric substations. The Resilient Digital Substation module (RDS) exceeds the shielding effectiveness test levels of the design basis. Following the development of the RDS module, efforts focused on real-time shielding compromise and event detection for EM threats. For threats, the HEMP/IEMI detector uses magnitude and frequency to determine the appropriate level of alert and need for operational response. Additionally, the detector reduces the need for technician field visits by regularly testing for shielding compromise, as well as providing forensic data for post-event analysis. The unique design features of the RDS module proved superior to more traditional designs with respect to HPEM mitigation and cost-effectiveness.

10.1 Financial comparative analysis

10.1.1 Introduction

The potential vulnerabilities of electric power systems to HPEM events have been documented by a number of research entities globally. Many of the design considerations, mitigation components, and construction methods have been enumerated in standards such as MIL-188-125-1 [1] and IEC 61000-5-10 [2] for fixed installations. Utility infrastructure can pose unique challenges to effective HPEM mitigation due to facility types and scale of the systems. While utility ownership and operational frameworks vary by region, electric delivery systems are often comprised of hundreds of individual substations. The scale of utility systems poses challenges from a cost perspective for both the initial installation and associated ongoing maintenance. Costs can be an impediment to HPEM mitigation efforts in power system infrastructure which led CenterPoint Energy to complete a comparative financial analysis of mitigation options from a complete system perspective.

[1]CenterPoint Energy Electric Services Corp., Houston TX, United States

10.1.2   Design basis

Before a financial comparative analysis could be completed, a design basis needed to be established and alternative designs developed. The primary electromagnetic threats to be considered were high-altitude electromagnetic pulse (HEMP), intentional electromagnetic interference (IEMI), and geomagnetic disturbances (GMD). The range of threats and hazards led to a number of functional requirements related to mitigation, detection, and testing. Of significance were the frequency ranges and radiated field levels expected to be experienced from intentional or natural sources. The radiated fields would then be used to determine the expected conducted energy.

Initial efforts led to a design that closely resembled the existing control houses but incorporated additional provisions including wave guides below cut-off, device power filters, surge suppression devices, circumferential bonding and grounding as well as shielded control cables. Legacy control houses range from approximately 3.5 m high by 6 m wide by 9 m deep to approximately 3.5 m high by 6 m wide by 26 m deep depending on the size of the substationas shown in Figure 10.1 [3].

The legacy-based design would use a shielded 6-sided enclosure with points of entry (POEs) addressed with electromagnetic compatibility (EMC) doors, appropriate modifications to the plumbing and climate control systems as well as the previously mentioned mitigations. Following the completion of the design based on legacy control houses, CenterPoint Energy developed a substation mitigation concept based on a compact modular design resulting in the Resilient Digital Substation (RDS) module [3] (Figure 10.2).

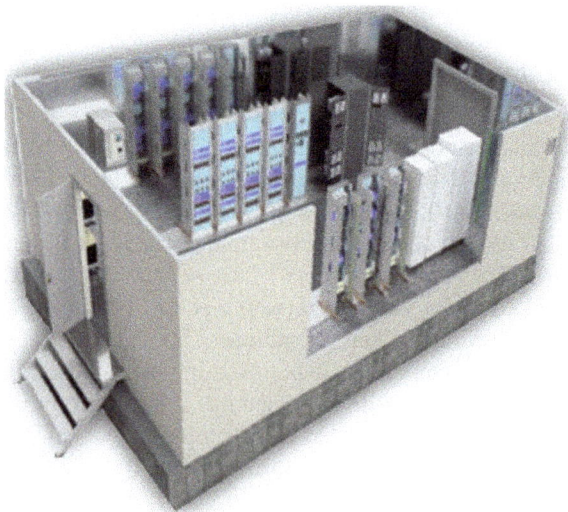

Figure 10.1   Rendering of a control house with the digital substation module alongside traditional relay panels

Figure 10.2   RDS module

The RDS module uses non-metallic reinforced fiber optic cable to eliminate the transfer of coupled energy from the substation yard to SCADA and relay equipment located in the control house. The use of fiber optics allowed for a significant reduction in the required shielded enclosure volume. The resulting design measures approximately 2 m high by 1.5 m wide by 1.5 m deep and is capable of fitting in an existing non-EMC control house (Figure 10.3) [4].

The photo shows the compact digital substation module in an existing control house at the end of the existing relay panels. An optional modification to the control house is the installation of taller doors; however, the module is able to fit in a standard door opening. The taller doors simplify the physical installation of the fully completed enclosure which has a total weight of approximately 700 kg. The compact design offered a number of technical advantages including a standardized HPEM mitigation design and hardware. These features eliminate the need to complete site-specific designs and allow the use of type testing to ensure compliance with performance requirements. Additionally, utilizing a separate compact module inside the legacy control house limits the number of door operations which can be a primary source of shield leakage and maintenance. The use of the existing control house avoids highly visible external cues highlighting which substations have been protected and which have not.

10.1.3   System comparison

With both designs complete, a comparison was conducted to rank the potential options based on their effectiveness in mitigating HPEM, initial construction costs, and ongoing maintenance costs. A flowchart was developed to assist in creating

Figure 10.3   RDS module installed in existing substation control house

mitigation options by considering the paths that both radiated and conducted energy would take to reach SCADA and relay equipment (Figure 10.4) [5].

Beginning with the free-field energy, a radiated field would directly interact with the control house and the equipment housed inside based on the level of attenuation inherent to the house. Existing control houses can be comprised of a number of different materials including brick, steel, or concrete. For scenarios involving IEMI, some protection may be provided by the line of side structures or fencing depending on the construction type and material used. Obstructions such as fencing were not considered for HEMP as the angle of incidence would render such a mitigation useless. The mitigation of HEMP will require the use of a 6-sided shielded enclosure with sufficient attenuation to protect the internally housed devices from radiated fields. The level of attenuation will depend on the electromagnetic source and design of the devices to be protected. Devices with composite or partially composite cases will require a higher level of attenuation for successful protection. The concept of successful protection must be determined based on the level of criticality of the device's function and tolerance to disruption, upset, and damage. For high-reliability systems, many seek to avoid disruption due to potential secondary effects. The most commonly used shielding approach is a metallic shield; however, the lack of a metallic floor in most existing control houses makes

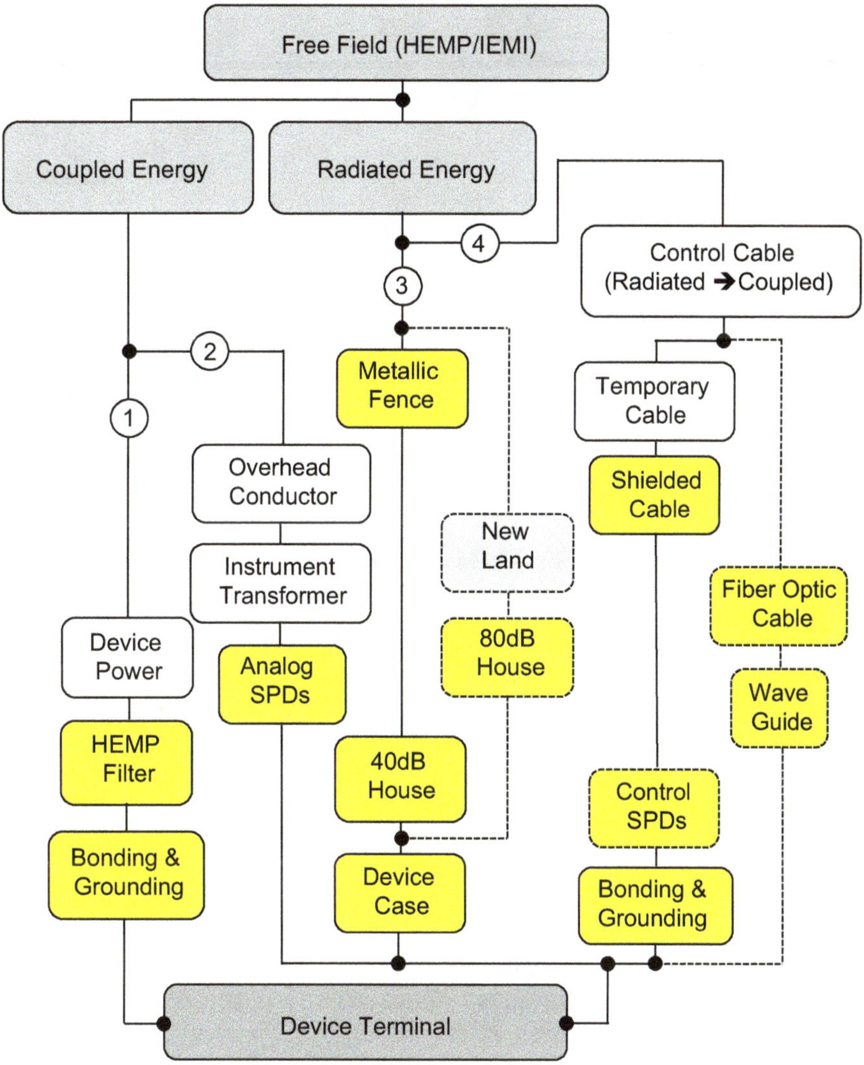

Figure 10.4   Mitigation flow chart

retrofit prohibitively expensive. In cases where HPEM mitigation is pursued, a common practice is to build a new control house based on a site-specific design to accommodate physical limitations and the needs of the substation. In some instances, the need to rebuild the control house may require additional property to allow an existing substation to remain in service while the HPEM mitigation project is under construction (Figure 10.5) [5].

Key design changes include managing POEs and the addition of a marshaling cabinet to terminate yard cables. For utilities that are not presently using shielded

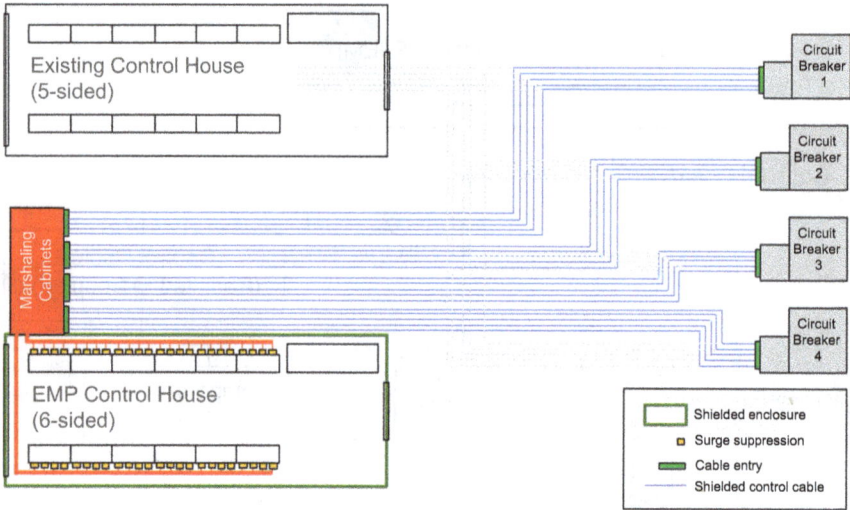

Figure 10.5   New 6-sided EMP shielded control house with marshaling cabinet, shielded metallic control cables, and cable entry points

cables, new control cables will need to be installed for all existing equipment. The type of cable shield is critical to shielding effectiveness and may require some existing shielded cables to be replaced; therefore, it is recommended to test an existing cable prior to estimating the mitigation costs to ensure the design assumptions are accurate. Cable replacement can add considerable costs to the project due to both materials and construction labor. Cable entry systems are used to properly bond and ground the cable shields of metallic cables entering the shielded volume of the control house. The cost will increase proportionally to the number of cables entering the house. The mitigation of transients should also consider reflections through instrument transformers connected to primary voltage power conductors. Energy coupled to overhead transmission lines can be mitigated through the use of surge protection devices (SPDs) connected to cables on the secondary voltage side of the instrument transformer. To achieve proper high-frequency grounding, the cable shield should be grounded on both ends which may introduce the potential for circulating currents on the shield of the conductor. Attention should be given to the substation ground grid at the time of installation and throughout the life cycle of the installation to minimize circulating currents. Final provisions include mitigations for the device power direct-current supply. External to the control house, consideration should be given to the station service which powers the battery charger and substation batteries. The incoming direct-current power supply should be protected by a HEMP filter.

The use of fiber optic cables will require the design to include fiber optic pass-throughs with provisions for changes in the propagation constant due to the fiber optic cable. The use of non-metallic reinforced fiber optic cable greatly reduces the number of metallic cables; however, the use of metallic cables is still required to provide device power to merging units located at the circuit breaker. The merging units serve as analog to digital converters and transmit information to the main relays in the control house. The significant reduction in metallic cables reduces the costs and shortens the construction time required. Ten legacy-based design options utilizing a traditional control house design with HPEM provisions were developed for comparison to the compact design of the RDS module (Table 10.1) [5].

All designs required filters, grounding and bonding, and analog SPDs. Key differences in costs resulted from the need for a new HPEM control house, temporary cable, SCADA and relay devices, the use of fencing, shielded cable, fiber optic cable, and the number of control cable SPDs required.

The need for a new control house can pose a number of challenges including specialized construction methods and increased maintenance once the control house is constructed. In urban installations, the additional property to rebuild the control house may be prohibitively expensive or unavailable for purchase. Land costs can add 50% to the project cost and delay the start of construction. Once construction is underway in an already operational substation, care must be taken to maintain the functionality of existing systems and avoid unintended outages which may impact electric consumers. The use of temporary control cables is often necessary during the rebuild process which adds complexity to both the design and construction phases of the project. This complexity increases the cost as

Table 10.1   Various option for achieving shielding effectiveness

Option #	Power	Coupled	Radiated	Radiated/Coupled
Option 1	Filter + Bonding and Grounding	Analog SPDs	N/A	Shield cable + Controls SPDs
Option 2	Filter + Bonding and Grounding	Analog SPDs	Device replacement + Fence	Shield cable + Controls SPDs
Option 3	Filter + Bonding and Grounding	Analog SPDs	30 dB House + New Relays	Temporary cable + shield cable + Controls SPDs
Option 4	Filter + Bonding and Grounding	Analog SPDs	80 dB House + New Relays	Temporary cable + shield cable + Controls SPDs
Option 5	Filter + Bonding and Grounding	Analog SPDs	Fence + 30 dB House + New Relays	Temporary cable + shield cable + Controls SPDs
Option 6	Filter + Bonding and Grounding	Analog SPDs	Device replacement	Fiber Optic Cable + Wave Guides
Option 7	Filter + Bonding and Grounding	Analog SPDs	Device replacement + fence	Fiber Optic Cable + Wave Guides
Option 8	Filter + Bonding and Grounding	Analog SPDs	30 dB House + New Relays	Fiber Optic Cable + Wave Guides
Option 9	Filter + Bonding and Grounding	Analog SPDs	80 dB House + New Relays	Fiber Optic Cable + Wave Guides
Option 10	Filter + Bonding and Grounding	Analog SPDs	Fence + 30 dB House + New Relays	Fiber Optic Cable + Wave Guides

Table 10.2   Cost options of the different solutions

Component	Component (cost) USD
HEMP Filter (each)	$10,000.00
House Bonding and Grounding (128 cables) (each)	$5,500.00
Yard Bonding and Grounding (128 cables) (each)	$600.00
Analog SPDs (each)	$150.00
Controls SPDs (each)	$150.00
Fence (1,000 ft @ $900/1ft)	$900,000.00
30 dB House (20 × 30 ft)	$200,000.00
80dB House (20 × 30 ft)	$700,000.00
Relay Panels (each)	$50,000.00
Temporary Cable (40k ft @ $1/ft)	$40,000.00
Shielded Cable (40k ft @ $3/ft)	$120,000.00
Fiber Optic Cable (8k ft @ $3/ft)	$24,000.00
House Fiber Optic scheme Bonding and Grounding (each)	$5,500.00

intermediate phases of the project must be engineered. There is also an increasing number of operational risks proportional to the number of phases required.

Costing of the different options was completed by determining the per unit cost for each component and applying them to a common substation design. For estimating purposes, a substation containing four transformers and 16 radial distribution feeders was selected. Each of the design basis criteria was applied to determine the total cost of initial construction. The costs used in this financial comparative analysis may vary based on the components selected and the time of quotation (Table 10.2) [5].

Labor times and costs will vary based on the final design; however, for this analysis, a labor rate of $60 USD per hour was used for all labor costs. For the substation used in this analysis, 16 legacy relay panels would be required versus a single RDS module. Designs based on shielded metallic cables will require 128 analog SPDs and 384 control SPDs.

Utilizing the compact modular design to mitigate HEMP effects simplifies design and construction processes due to standardized hardware. A digital protection system can be reconfigured to a wide range of substation topologies with programming changes only. A standardized relay design also allows for units to be ordered in advance of a specific project and stored in a warehouse until the project is initiated. Additionally, type testing of the HEMP mitigations using both radiated and conducted energy testing decreases the test requirements at the time of construction. Each component of the module was tested to ensure the efficacy of the complete system [4]. A fiber optic-based design eliminates the need to utilize temporary cables which reduces construction time from months to days. Typical construction times for shielded metallic cable-based designs are approximately 12 months versus 15 days for fiber optic-based designs (Figure 10.6) [5].

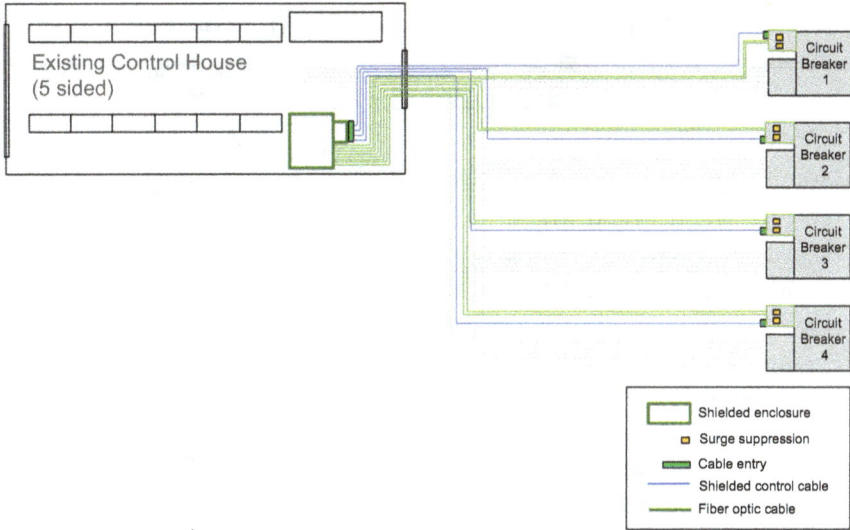

Figure 10.6   Digital substation module installed in an existing control house using fiber optic control cables and shielded power cables

The ability to reuse the existing control house eliminates the potential need to purchase additional property. Construction is completed by first installing the merging units at each one of the circuit breakers and only requires a single breaker outage in sequential order. Following the installation of the merging units no additional outages are required to complete the HEMP protection of the entire substation. The fiber-optic cable does not interpose with the existing control cables and the consolidation of signals typically carried over multiple metallic cables is accomplished with a single fiber-optic cable. As a result, the number of SPDs and associated costs is greatly reduced. Shielded metallic device power cables are utilized; however, the significant reduction in metallic cables reduces the complexity of cable entry systems.

In order to compare the value of different designs, consideration needed to be given to the effectiveness of the designs to mitigate HPEM impacts. The only design subjected to empirical testing was the RDS module. Due to its dimensions, the unit was able to fit in the test volume of an RS 105 HEMP simulator and tested to 50 kV/m using an unclassified E1 waveform [6]. Additionally, conducted energy testing was completed based on expected coupling effects. The effectiveness of all other designs was developed based on modeling and simulation. The mitigation effectiveness of fencing assumed that a proper line of sight assessment was made to determine the appropriate fence height to prevent exposure from a localized IEMI attack. These studies should assume the field strength and wavelength of the IEMI source as well as the topological features of the land surrounding the substation. It was assumed that the perpetrator would not utilize devices to place the IEMI source above the fence level. The sum of these differences results in the initial cost of the module ranging from 1/3 to 1/5 of equivalent designs. Designs with a green bar

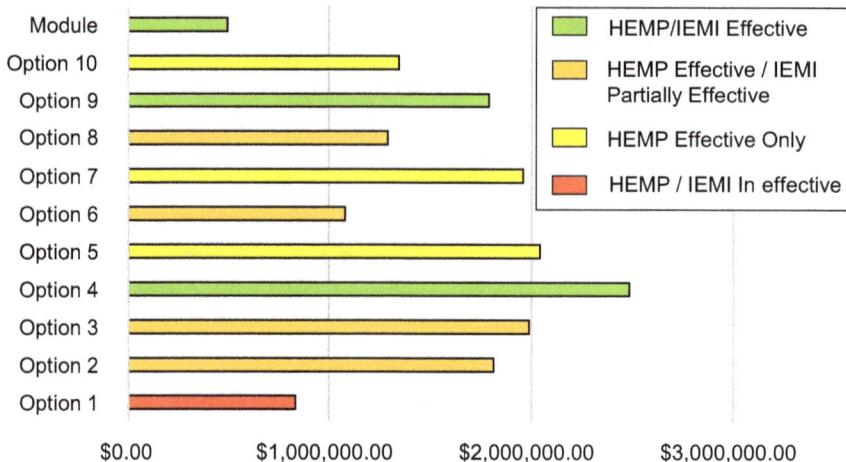

Figure 10.7   Design considerations for the module and recurring testing

were considered effective in mitigating HEMP and IEMI (options 9 and 4), orange bars mitigate HEMP only (options 2, 3, 6, and 8), yellow indicates HEMP and partial IEMI (options 5, 7, and 10), and option 1 shown in red was deemed ineffective against HEMP and IEMI (Figure 10.7) [5].

10.1.4   Strategy development

As CenterPoint Energy considered the deployment of the RDS module technology systemwide, several factors played a role in the development of a strategy. One of the most important factors is scalability, for the program to be successful the overall number of design changes needed to be minimized. After evaluating all substations, groupings of similar substation topologies were identified to add efficiency to the relay programming process and establish a base hardware configuration that was universally applicable. This allows for the module hardware to be designed for the ultimate substation design (4 transformers and 16 feeders) followed by site-specific programming of the necessary relays at the time of construction. Should the substation be expanded in the future; the additional work will be limited to hardware external to the RDS module and additional programming without the need to add additional hardware inside the RDS module. Once the deployment strategy was understood the focus shifted to assessment of supply chain partners to ensure the necessary components would be available. To accommodate the manufacturing process, a multi-phase plan was developed to avoid supply chain constraints and establish a collaborative project schedule that could be managed by all stakeholders. The RDS modular design provides inherent engineering and construction time savings throughout the program. This allows reduced outage time per project and decreased operational risks. During the development of the deployment strategy, opportunities to reduce operational and maintenance costs associated with the deployment of the RDS module were identified.

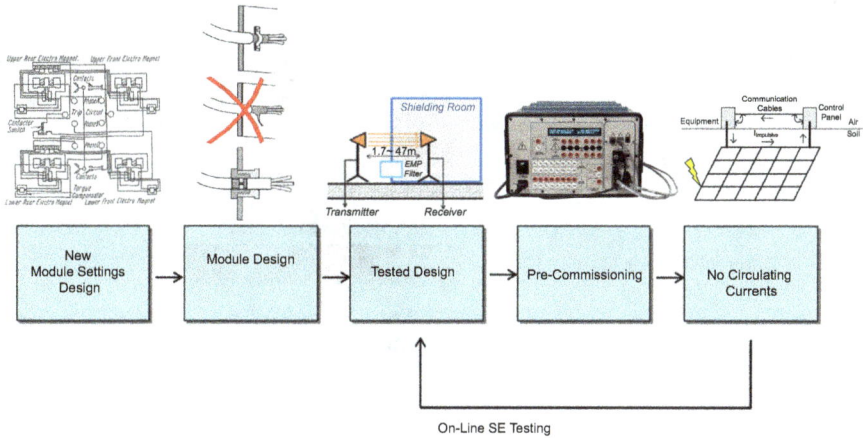

Figure 10.8   Design considerations for the module and recurring testing

10.1.5   Maintenance cost

The digital substation module has the potential to reduce the overall maintenance costs as compared to traditional protection and control (P&C) panels. The module has minimal components that need maintenance. The overall design will only require periodic inspections of the shielded cable to identify any degradation of the cable as well as cleaning of waveguides. The shielding effectiveness will be tested through the use of the integrated online HEMP/IEMI detector and shielding compromise solution (Figure 10.8) [4].

The use of an online shielding compromise solution will aid in reducing maintenance costs by eliminating the need for on-site testing annually. Additionally, the surge suppression devices have the ability to alarm for failures limiting the need for field inspections and providing continuous monitoring to ensure operational readiness. When evaluated over a 30-year period for net present value (NPV), the maintenance costs of the compact digital module are significantly less than alternative designs (Figure 10.9) [5].

In addition to the reduced inspection costs, maintenance cost reductions also result from a reduced number of surge suppression devices. The use of fiber optic-based controls significantly reduces the number of surge suppression devices required. The reduction in surge suppression devices results in a lower forecasted replacement quantity when using the same expected failure rate for each design.

CenterPoint Energy set out to identify an effective, cost-efficient, and scalable HPEM mitigation solution for deployment at electric substations. The ability to utilize a standardized type-tested mitigation design optimizes deployment strategy options when considering a system-wide deployment. A standardized design also streamlines the training process for both engineering staff and field technicians. The initial deployment of RDSs allows CNP to ensure HPEM resiliency using a deployment strategy with low operational risk. Initial deployments disable controls from the RDS for day-to-day operations but continue to allow performance data

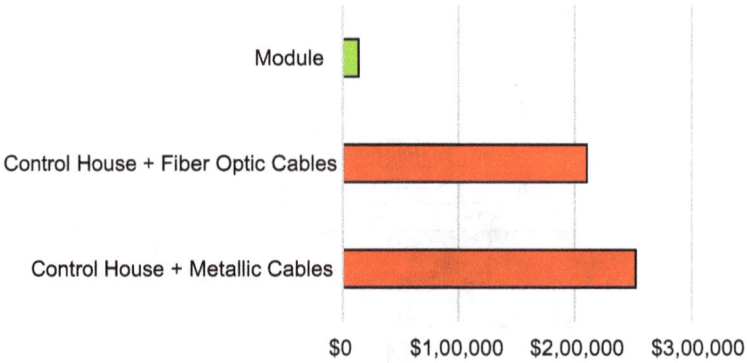

Figure 10.9   NPV of life cycle maintenance costs per site

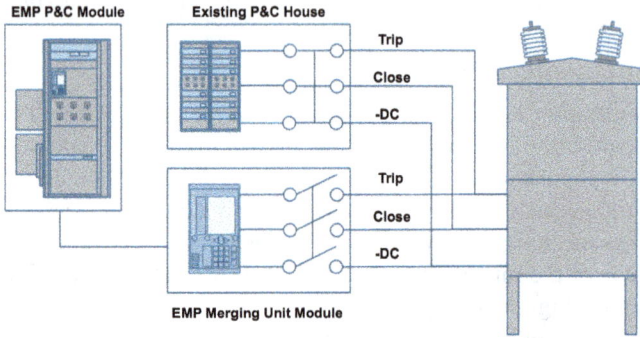

Figure 10.10   Design considerations for the module and recurring testing

collection for both shielding effectiveness from HPEM mitigations and protective relay operations for faults on the power system (Figure 10.10) [5].

10.1.6   Summary and limitations

The initial installation allows for the RDS module to operate in parallel with the existing P&C equipment. Should an HPEM event damage existing P&C equipment, the HPEM-hardened P&C in the RDS module can be quickly activated by closing the open switches at the substation. The RDS has met all identified tests for HPEM mitigation and cost efficiency leading to its adoption by CenterPoint Energy for systemwide deployment.

With multiple RDS modules in service, our next efforts will focus on long-term performance and maintenance data collection. To date, 3 years of performance data have been collected and units have only required minimal maintenance. The continued collection of maintenance data will assist in validating the assumptions utilized in the financial comparative analysis.

References

[1]   Mil-Std-188-125-1, "High-Altitude Electromagnetic Pulse (HEMP) Protection for Ground-Based C4I Facilities Performing Critical, Time-Urgent Missions," Part 1, Fixed Facilities, 2005.

[2]   IEC/TS 61000-5-10 Ed. 1.0 (2017-05-18), "Electromagnetic Compatibility (EMC) – Part 5-10: Installation and Mitigation Guidelines – Guidance on the Protection of Facilities against HEMP and IEMI," Basic EMC publication.

[3]   E. Easton and K. Bryant, AMEREM 2018, "Practical Application of EM Mitigation for Critical Infrastructures," Santa Barbra California, USA, August 2018.

[4]   E. Easton, K. Bryant and W. Radasky, "Testing of a Module for Electrical Substations to Demonstrate HEMP and IEMI Protection and GIC Detection," 2020 *IEEE International Symposium on Electromagnetic Compatibility & Signal/Power Integrity (EMCSI)*, Reno, NV, USA, 2020, pp. 442–447, doi:10.1109/EMCSI38923.2020.9191506.

[5]   E. Easton and C. Wafo, GLOBALEM 2022, "Financial Comparative Analysis of Substation EMP Mitigation Approaches," Abu Dhabi, UAE, November 2022.

[6]   E. Easton, R. Horton, K. Bryant and J. Butterfield, "Assessment of EMP Hardended Substation Protection and Control Module," 2020 *IEEE International Symposium on Electromagnetic Compatibility & Signal/Power Integrity (EMCSI)*, Reno, NV, USA, 2020, pp. 448–453, doi:10.1109/EMCSI38923.2020.9191681.

Chapter 11

Three-dimensional FDTD-based lightning transient analysis of secondary circuits with shielded control cables over grounding structures in a substation

Akiyoshi Tatematsu[1] and Akifumi Yamanaka[1]

Nowadays, numerous sensitive electronic devices are utilized in low-voltage circuits, that is, secondary circuits of substations. When transmission lines (TLs) and substations are struck by lightning, lightning currents enter the primary circuits and the grounding structure of the substation through the occurrence of back-flashover at TL towers, the operation of surge arresters installed in the substation, and so forth, which causes the electromagnetic coupling and the transient ground potential rises (GPRs) of the grounding structure, inducing voltages in the secondary circuits. Shielded control cables are one of the typical countermeasures for protecting these secondary circuits from electromagnetic disturbances. Therefore, accurately predicting electromagnetic transient phenomena in secondary circuits and assessing the effectiveness of shielded control cables are crucial for designing appropriate countermeasures. The finite-difference time-domain (FDTD) method, which is one of the numerical electromagnetic computation methods, has recently been widely adopted for the electromagnetic transient analysis of three-dimensional (3-D) and grounding structures. This study utilizes an FDTD-based transient analysis technique to represent shielded control cables and simulates the voltages induced on a multi-core shielded control cable positioned over grounding structures during a lightning impulse current injection. The simulated results are then compared with measured waveforms for validation.

11.1 Introduction

Lightning strikes and the operation of disconnectors and circuit breakers generate lightning and switching transients, which may induce electromagnetic disturbances in secondary circuits equipped with numerous sensitive electronic devices nowadays, in substations and power plants [1]. Accurate prediction of electromagnetic

[1]Central Research Institute of Electric Power Industry, Kanagawa, Japan

transient phenomena and assessment of the effectiveness of countermeasures through numerical simulations are essential for designing protection measures. Traditionally, the analysis of electromagnetic transients in secondary circuits has been conducted using simulation techniques based on circuit theory or TL theory [2–10].

In Japan, a survey covering approximately 10 years up to 1999 investigated faults and malfunctions of panel-type equipment that occurred in substations, power plants, and other electric power facilities. The survey revealed that the cause of 72% of disturbances was presumed to be lightning transients [11,12], and thus it is essential to develop techniques for analyzing lightning transient phenomena in secondary circuits. Nowadays, the application of full-wave numerical approaches such as the finite-difference time-domain (FDTD) method [13], the method of moments [14], and the finite-element method [15], has become very useful, and they have been widely and successfully employed for electromagnetic transient analysis of 3-D structures such as TL towers and buildings and grounding systems such grounding grids and earth electrodes [16]. Among the full-wave numerical approaches, the FDTD method is advantageous in terms of its capability of simulating inhomogeneous soil parameters, nonflat ground surfaces such as mountains, and nonlinear phenomena such as the voltage–current relationship of surge arresters [17,18] and flashover across arcing horns at TL towers [19,20]. Thanks to the development of new simulation techniques, for example, to represent thin wires for FDTD-based electromagnetic transient simulations and the proliferation of high-performance computers, for example, with graphical processing units (GPUs), the FDTD method has become a very powerful tool for analyzing electromagnetic transient phenomena in power plants, substations, TLs, distribution lines, and other electric power facilities [21–28].

When lightning strikes TLs and substation, lightning transients enter the substation, that is, the primary (main) circuit and the grounding structure of the substation, which induces electromagnetic disturbances in the secondary circuit through the following phenomena: (i) surge transition; surges transferred from the primary to secondary circuits at instrument transformers such as voltage and current transformers (VTs and CTs), (ii) transient GPRs of the grounding structure and electromagnetic coupling due to lightning transients propagating through the grounding structure, (iii) electromagnetic and electrostatic coupling directly from the lightning transients propagating through the primary circuit, and so forth [11]. In [29], the 3-D FDTD method was applied to the transient analysis of voltages induced in secondary circuits when transient voltages entered the primary circuit of a test platform simulating a substation. The test platform set up in Akagi Testing Center of Central Research Institute of Electric Power Industry (CRIEPI) included two grounding grids, a gas-insulated switchgear (GIS) model, and a digital-type protection relay, where the protection relay was connected to the secondary circuit of VT and CT using a control cable. The study proposed techniques for incorporating the frequency-dependent surge transition effect at VT and CT and for simulating the secondary circuit using a hybrid approach combining the 3-D FDTD method and TL theory. The FDTD simulations were validated through comparison with the measurements using the test platform. Recently, [30] introduced a technique for modeling shielded control cables, which can account for the effect of the frequency-dependent surface transfer impedance of shields of control cables in 3-D FDTD-based transient simulations. Using the proposed technique, the study calculated voltages induced on both unshielded and shielded control cables in the test platform, when a lightning impulse current was injected into the grounding grid, with calculated results validated against measured waveforms.

In this study, we employed the technique for representing shielded control cables to perform FDTD simulations of a shielded control cable with a tape-type shield, which is commonly used in extra-high-voltage and high-voltage substations for protecting secondary circuits. First, we provided an overview of the modeling of shielded control cables within the 3-D FDTD framework. Next, we calculated voltages induced on the shielded control cable, which was placed over remote grounding structures, considering the surface transfer impedance of the multi-core shielded control cable, for a lightning impulse current injected into the grounding structure. Lastly, we modeled the test platform representing the primary and secondary circuits of a substation to compute voltages induced in the secondary circuit at the protection relay due to a lightning impulse current injected into the grounding grid. This was done to evaluate the effectiveness of the shielded control cable and surge-absorbing capacitors, which are commonly installed at protection relays in substations. The results from the 3-D FDTD simulations were validated by comparing them with measurements.

11.2   Modeling of shielded control cables

There are several techniques developed for representing different types of cables in FDTD-based transient simulations, including coaxial cables [31–33], pipe-type power cables [34], and coaxial cables with multiple conductive layers [35]. These techniques have been applied to analyze electromagnetic transients in various scenarios, such as power cables in a substation [25], power cables in a distribution line [34,35], and a TL tower with a power cable [36]. In this study, we utilized the technique specifically for representing multicore shielded control cables in FDTD-based transient analysis [30]. An overview of this technique is given below.

In this technique, a lossy thin wire, which can take into account the frequency-dependent skin effect of a tubular conductor [31], is placed directly in a 3-D FDTD analysis space to simulate the sheath of a shielded control cable. Note that the thin-wire representation technique in [31] can be applied to both tubular and solid conductors. On the other hand, electromagnetic transient phenomena inside the sheath are solved by applying the 1-D TL theory. Assuming that (i) the surface transfer impedance is the dominant effect while the surface transfer admittance can be ignored, and (ii) the transient phenomena are in the transverse electromagnetic mode, voltages and currents induced on the shielded control cable (N cores) are solved by the following expressions on the basis of 1-D TL theory and weak coupling [37,38]:

$$\frac{\partial \mathbf{V}(l,t)}{\partial l} + \int_0^t \zeta_c(t-\tau)\frac{\partial \mathbf{I}(l,t)}{\partial \tau}d\tau + L'\frac{\partial \mathbf{I}(l,t)}{\partial t} = \mathbf{Z_t} * I_s(l,t), \tag{11.1}$$

$$\frac{\partial \mathbf{I}(l,t)}{\partial l} + C'\frac{\partial \mathbf{V}(l,t)}{\partial t} = 0. \tag{11.2}$$

$V(l, t)$, $I(l, t)$, and $I_s(l, t)$ are voltages and currents along the N-cores, and sheath current, respectively. L', C', and Z_t are, respectively, the per-unit-length (p.u.l.) inductance and capacitance matrices, and the surface transfer impedance of the control cable. As shown in [30], shielded control cables used in substations have frequency-dependent surface transfer impedances. To take into account such frequency-dependent effects, the convolution technique is applied to calculate the right-hand side of (11.1). The output of the right-hand side of (11.1) $v(s)$ in the frequency domain is expressed as

$$v(s) = Z_t(s)I_s(s) \ (s = j\omega = j\,2\pi f). \tag{11.3}$$

On the assumption that $v(s)$ tends to zero ($f \to \infty$), the surface transfer impedance can be approximated:

$$Z_t(s) \cong \sum_{k=1}^{K} \frac{R_k}{s - p_k}, \tag{11.4}$$

where p_k and R_k are the kth pole and the residue of p_k, respectively. The right-hand side of (11.1) can be calculated in the time domain using the trapezoidal rule and the recursive convolution technique [39].

As mentioned above, a lossy thin wire placed in a 3-D FDTD analysis space simulates the external conductor sheath of the control cable, and then the current I_s flowing through the sheath included in (11.1) can be obtained from the rotational magnetic fields around the thin wire calculated by the 3-D FDTD method using Ampere's law. Then, the 1-D FDTD method is applied to solve (11.1) and (11.2) and calculate the voltages and currents on the N cores, where $I_s(l, t)$ in (11.1) is derived from I_s calculated by the 3-D FDTD method using the first-order basis function for spatial representation and the second-order basis function for temporal representation. In actuality, control cables installed in secondary circuits of substations have a twisted structure, but in this technique, the cores of the shielded control cables are approximately treated as parallel-wire pairs and L' and C' are assumed to be uniform along the control cables because the layout of the cores and sheath in a cross-section is almost the same along the control cables.

11.3 Voltages induced on shielded control cable over remote grounding structures

In some power plants and substations, shielded control cables are placed over remote grounding structures, where the grounding structures are connected using grounding wires. When the sheaths of the shielded control cables are grounded at both ends, a lightning current entering grounding structures flows also into the cable sheaths, inducing voltages on the control cables. In this study, we performed 3-D FDTD simulations to calculate voltages induced on a shielded control cable with a tape-type shield grounded at both ends, positioned over remote grounding structures for a lightning impulse current injected into the grounding structure. The FDTD results were compared with measured waveforms for validation.

11.3.1 Experimental setup

First, using an experimental setup composed of a grounding grid, a ring earth electrode, and a grounding wire between them, we measured the GPRs of the grounding structures. Figure 11.1 shows the top view of the experimental setup. The sizes of the grounding grid and ring earth electrode were 30 m × 60 m and 4.5 m × 9 m, respectively. The grounding grid was connected to the ring earth electrode by the grounding wire. The grounding grid, ring earth electrode, and grounding wire were composed of bare copper wires (cross-section: 60 mm^2), and they were, respectively, buried 0.5 m, 0.4 m, and 0.5 m below the ground surface. A deep earth electrode with a length of 65 m was buried 2 m away from the ring earth electrode, and it was connected to the ring earth electrode by a grounding wire (cross-section: 60 mm^2) placed 0.5 m below the ground surface. A pulse generator to inject a current was placed close to the grounding grid, and it was connected to the grounding grid by a wire. A current injection wire (cross-section: 2 mm^2) at a

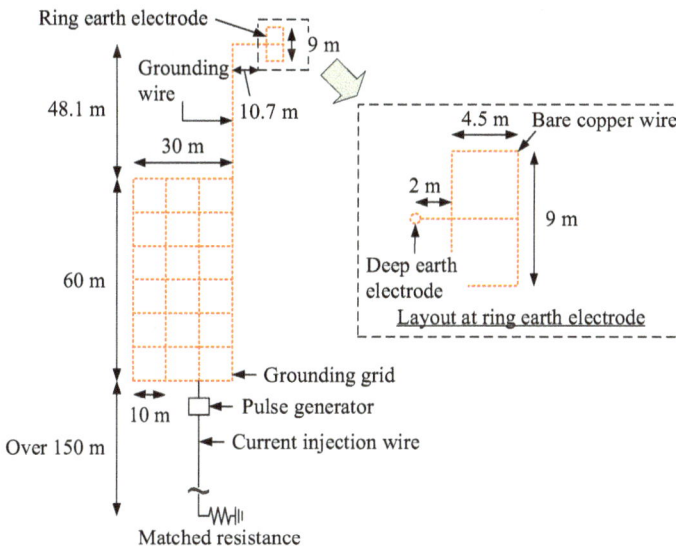

Figure 11.1   Experimental setup composed of remote grounding structures

height of 0.5 m was connected to the pulse generator. The far end of the current injection wire was grounded through its matched resistance. A lightning impulse current was injected into the grounding grid with the pulse generator, and the GPRs of the grounding grid and ring earth electrode were measured at positions A-1, A-2, A-3, and B-1 as shown in Figure 11.2. The GPRs were measured as voltage differences between the grounding structures and a voltage reference wire (cross-section: 2 mm^2) placed at a height of 1.5 m. The far end of the voltage reference wire was grounded by an earth electrode through its matched resistance to avoid the occurrence of reflected waves. To measure the GPRs, the voltage reference wire was extended to positions A-1, A-2, A-3, and B-1 as indicated by dotted lines in Figure 11.2.

Second, as shown in Figure 11.3, a shielded control cable with a tape-type shield and two cores (cross-section: 3.5 mm^2) was placed on the soil surface over the grounding grid and the ring earth electrode, and the cable sheath was grounded at both ends. A resistor of 100 Ω was connected to the two cores at each end. As shown in Figure 11.3, one of the cores was connected to the ring earth electrode. We measured the voltage induced on the shielded control cable, that is, the voltage across the resistor at the far end of the control cable and the sheath current on the grounding wire of the sheath at position B-2 for the lightning impulse current injection.

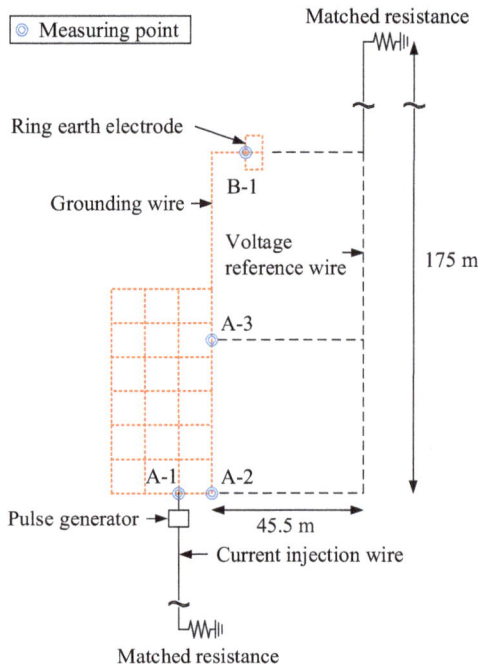

Figure 11.2   Layout of voltage reference wire used for measuring GPRs

11.3.2   Calculation model

Figure 11.4 shows a 3-D FDTD analysis space with dimensions of 580 m ×
475.5 m × 467.5 m simulating the experimental setup, where the soil surface was
positioned at a height of 266 m. All the surfaces were treated as absorbing
boundary conditions of the second-order Liao [40] to assume an open space. In
the analysis space, the grounding grid, ring earth electrode, and grounding wire

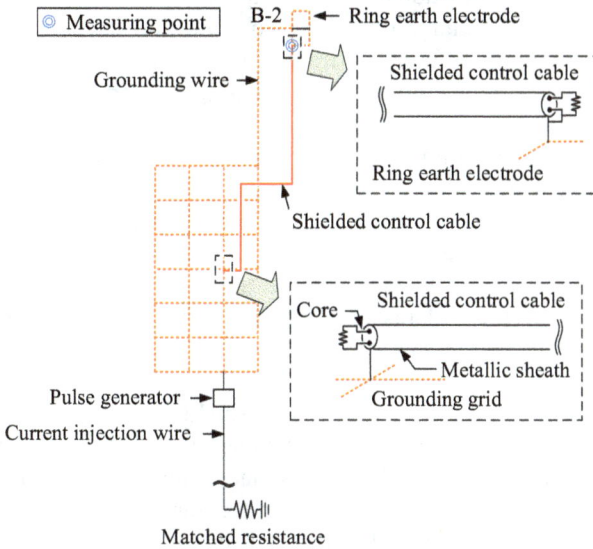

Figure 11.3   Layout of shielded control cable

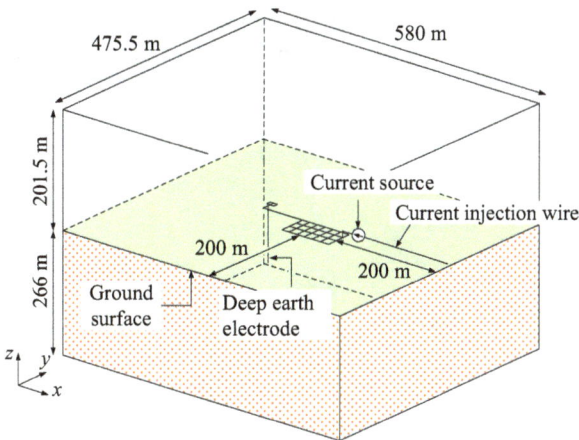

Figure 11.4   Analysis space modeling experimental setup in Figure 11.1

between them were simulated thin wires with a radius of 4.4 mm using the thin-wire representation techniques [41,42]. The deep earth electrode, current injection wire, and voltage reference wire were also simulated by thin wires, the radii of which were 33 mm, 0.8 mm, and 0.8 mm, respectively. The far end of the thin wire representing the current injection wire was directly attached to the side surface of the analysis space simply to simulate its matched condition. On the other hand, for accurately assessing the GPRs of the grounding structures, it is crucial to simulate the positional relationship between the grounding structures and the far end of the voltage reference wire, and the far end of the thin wire representing the voltage reference wire was located at the same position as the experimental setup, where a thin wire representing the earth electrode was buried and a matched resistance [43] was inserted between the thin wire representing the earth electrode and the thin wire representing the voltage reference wire to simulate the matched condition. A lumped current source was connected to the grounding grid model to inject a current, and the output waveform of the current source was the same as the measured injected current. The relative permittivity of the soil was set uniformly to 30 on the basis of the results of the previous study [28], whereas the soil resistivity was assumed to have three layers, the resistivities of which were set as given in Figure 11.5, on the basis of the previous studies [28,44].

To calculate the voltage induced on the shielded control cable, the shielded control cable was simulated using the technique presented in Section 11.2, and a thin wire representing the cable sheath was placed approximately 0.2 m above the ground surface. The frequency-dependent surface transfer impedance of the control cable was the same as the ones studied in [30] because the type and manufacturer of the control cable used in the aforementioned measurement were the same as those used in the previous study.

In [30], the frequency characteristics of the surface transfer impedance of a shielded control cable were measured. In the measurement, a 5 m shielded control cable was mounted horizontally over a grounded large copper plate. The control cable had two cores, labeled as cores #1 and #2. To inject a current into the cable

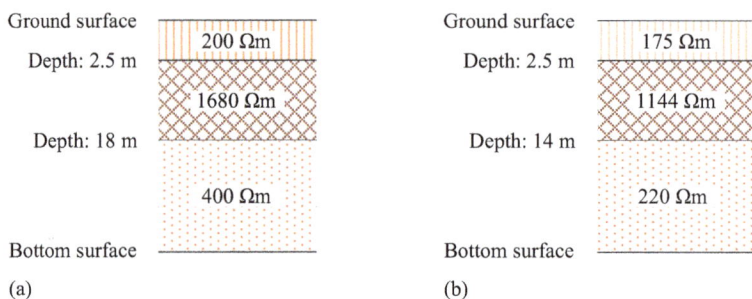

Figure 11.5   *Soil resistivity used in 3-D FDTD model. (a) On the side of the grounding grid. (b) On the side of the ring earth electrode.*

sheath, a pulse generator was connected to the sheath at one of the control cables. At this end, both cores were connected to the sheath. At the other end of the cable, the sheath was connected to the copper plate through a resistor. When a current with frequencies ranging from 10 kHz to 5 MHz was injected into the cable sheath using the pulse generator, the sheath current and the voltage between the core and sheath at the far end were measured, and then the frequency-dependent surface transfer impedance was obtained, as shown below. In addition, voltages induced on the cores due to the current injected into the cable sheath for the same frequency range were measured using a similar experimental setup with the 5-m control cable, where a resistor was inserted between the two cores at both ends and one of the cores was connected to the shield at one end, which simulated a realistic scenario where one of the cable cores is grounded in the secondary circuits of VT and CT circuits in substations. From the measured results, the surface transfer impedance $Z_{t1}(f)$ for core #1 was obtained by:

$$Z_{t1}(f) = \frac{F(V_{c1}(t))}{F(I_s(t)) \cdot L},$$
(11.5)

where $V_{c1}(t)$, $I_s(t)$, $F(\cdot)$, and L denote the voltage between core #1 and the sheath, the sheath current, the Fourier transform, and the cable length, respectively. Figure 11.6 shows the measured result of the surface transfer impedance for case #1 and the one reproduced by the FDTD method simulating the above experimental setup. In the FDTD simulations, the surface transfer impedance for core #1 $Z_{t1}(f)$ was simulated on the basis of the measured result. On the other hand, to incorporate the effect of slight unbalances between the surface transfer impedances of cores #1 and #2 on the induced voltages on the 5-m control, the amplitude of the surface transfer impedance for core #2 Z_{t2} was given using the following expression on the basis of the measured surface transfer impedance for core #1 (Z_{t1}) on the assumption that the phase of Z_{t2} was the same as that of Z_{t1}:

$$|Z_{t2}(f)| = k \times |Z_{t1}(f)|.$$
(11.6)

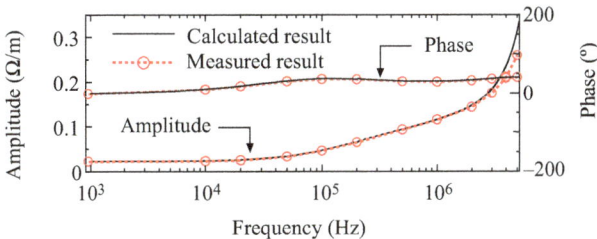

Figure 11.6   Measured and calculated surface transfer impedance of the control cable for core #1 [30]

A proper value of k was estimated to be 1.03 at frequencies of 1 to 50 kHz and to be 1.0 at higher frequencies by comparing the calculated induced voltages with the measured results.

Similar to the previous study [30], L' in (11.1) was computed using a conductor-subdivision-based method for calculating the series impedance matrix of conductors with arbitrary cross sections [45–47]. In this method, the cross sections of the cores and sheath of the shielded cable were divided into small square sub-conductors. On the other hand, C' in (11.2) was calculated through the electrostatic field analysis using the two-dimensional surface charge method [48], which is a boundary-dividing method similar to the boundary element method. The electro-static field analysis code can model conductors and dielectrics by dividing their surfaces into curved line elements.

The aforementioned 3-D FDTD analysis space was divided into nonuniform cells, the sizes of which ranged from 0.1 m to 2 m, whereas the time step was set to 96.3 ps. The total numbers of the cells used for calculating the GPRs and the induced voltages were, respectively, $641 \times 939 \times 312$ and $590 \times 853 \times 312$. The FDTD simulations were performed on a GPU-based parallel computer to reduce the calculation time dramatically.

11.3.3   Calculated and measured results

Figure 11.7(a) shows the calculated and measured currents injected into the grounding grid at position A-1. The peak values of both calculated and measured

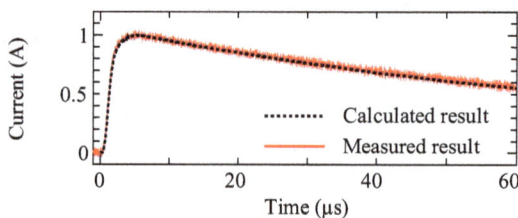

(a)

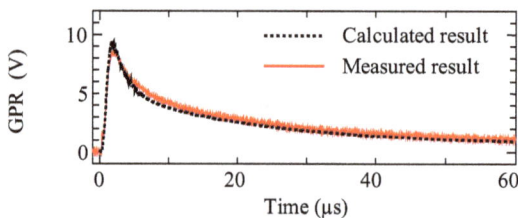

(b)

Figure 11.7   Calculated and measured results at position A-1. (a) Current and (b) GPR.

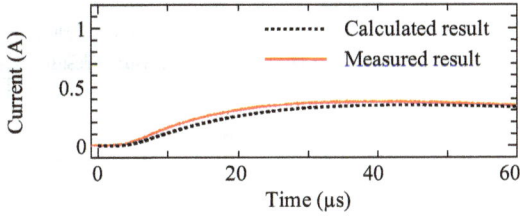

(a)

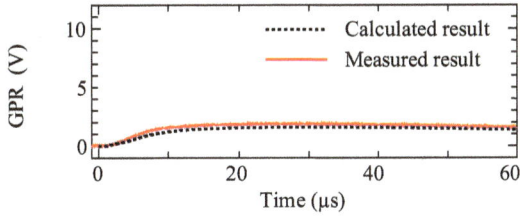

(b)

Figure 11.8   Calculated and measured results at position B-1. (a) Current and
(b) GPR.

injected currents were scaled to 1 A. The rise time (time interval between 10% and
90% of the peak pulse amplitude divided by 0.8) and the pulse width (time interval
between the points on the leading and trailing edges of a pulse at which the
instantaneous value is 50% of the peak pulse amplitude) of the injected current
were 2.1 μs and 68 μs, respectively. Figure 11.7(b) shows the calculated and
measured GPRs at position A-1. Figure 11.8(a) and (b) shows the calculated and
measured currents flowing through the grounding wire connected to the ring earth
electrode and the GPR at position B-1, respectively. The difference between the
peak values of the calculated and measured currents at position B-1 is 7.9%. The
discrepancies between the calculated and measured peak GPRs at positions A-1 and
B-1 are less than 14%. Furthermore, the calculated results of the currents and GPRs
at positions A-2 and A-3 are also in good agreement with the measured waveforms.
These findings confirm that the 3-D FDTD method accurately reproduces electromagnetic transient phenomena in the grounding structures. Figure 11.9 shows the
calculated and measured results of the current flowing through the grounding wire
of the cable sheath and the induced voltage across the resistor inserted between the
two cores at position B-2. Figure 11.9 demonstrates a good agreement between the
calculated and measured sheath current and induced voltage on the shielded control
cable, with differences in the peak values being 3.9% and 10%, respectively. This
further validates the applicability of the 3-D FDTD method to the electromagnetic
transient analysis of shielded control cables placed over remote grounding
structures.

(a)

(b)

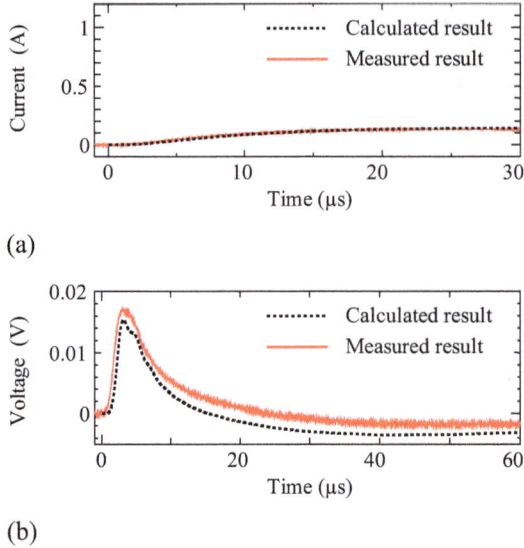

Figure 11.9   Calculated and measured results at position B-2. (a) Sheath current and (b) induced voltage.

11.4   Voltages induced in secondary circuits over grounding structures

Here, using the aforementioned 3-D FDTD technique, we modeled a test platform simulating primary and secondary circuits of a substation, calculated the voltages induced in the secondary circuit for a lightning impulse current injected into a grounding grid, and validated the calculated results against measured data.

11.4.1   Experimental setup and FDTD model

Figure 11.10 shows an experimental setup of the test platform to simulate primary and secondary circuits of a substation to measure voltages induced in the secondary circuits. The test platform included two grounding grids, a GIS model, and a digital-type protection relay. Detailed information about the test platform can be found in [30], and its outline is provided below. The GIS model consisted of a gas-insulated bus model, a VT, and a CT. The external sheath of the GIS model was grounded at multipoints. The protection relay was installed in a control building. As shown in Figure 11.10, an unshielded or shielded control cable with two cores (cross-section: 3.5 mm^2) was placed above the grounding grids, connecting the secondary circuit of the CT to the protection relay. The type of the shielded control cable was the same as that used in Section 11.3. As shown in Figure 11.10, one of the cores of the unshielded or shielded control cable placed on the soil surface was connected to the grounding terminal of the protection relay to simulate a real scenario in a substation. The VT, CT, and protection relay were the actual equipment

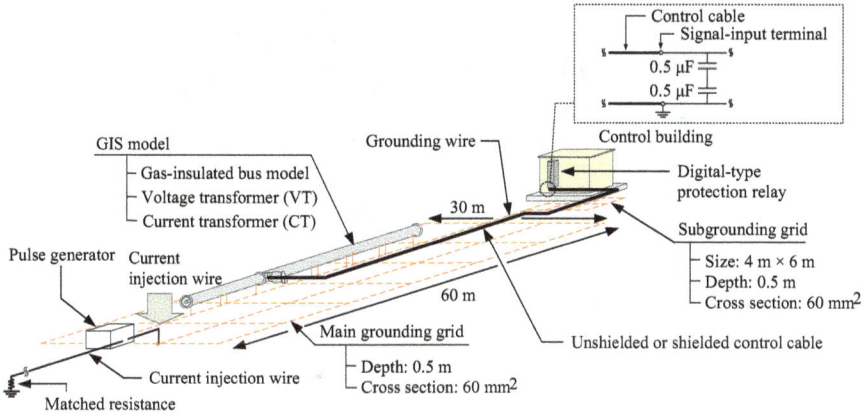

Figure 11.10 *Experimental setup used for measuring voltages induced on a control cable in the secondary circuit of a test platform*

for a 66-kV voltage class. The protection relay was equipped with a surge-absorbing capacitor of 0.25 µF at the signal-input terminal as shown in Figure 11.10. In this study, to evaluate the effectiveness of the surge-absorbing capacitor in addition to the sheath of the control cable against induced voltages, we injected a lightning impulse current into the grounding grid with the same pulse generator as that used in Section 11.3, we measured voltages induced on the control cable at the protection relay with and without the surge-absorbing capacitor for both unshielded and shielded control cables.

We carried out 3-D FDTD simulations of the above test platform. The dimensions of the 3-D FDTD analysis space were 330 m × 380 m × 300 m, where the soil surface was positioned at a height of 150 m. The electrical parameters of the soil were the same as those used in the FDTD model presented in Section 11.3.2 (see Figure 11.5(b)). All the external surfaces were treated as absorbing boundaries of the second-order Liao's technique. The other details of the calculation model can also be described as in [30], and we present an outline of the control cable model here. The shielded control cable was simulated using the technique in Section 11.2, and the same parameters as those used in Section 11.3 were applied for modeling the shielded control cable, where the height of the thin wire representing the cable sheath was set to 0.2 m approximately. On the other hand, the unshielded control cable with a twisted-wire structure was modeled by a twisted-wire pair using the hybrid approach of the FDTD method and TL theory [49]. In this technique, Agrawal *et al.* model was employed to calculate voltages and currents induced on a twisted-wire pair exposed to an electromagnetic field. In the Agrawal *et al.* model [50], it is necessary to incorporate exciting electric fields into TL-theory-based equations, and the electric fields along the conductors in the twisted-wire pair in the absence of the conductors are set by applying an interpolation technique to the electric fields obtained by the 3-D FDTD simulations using the aforementioned

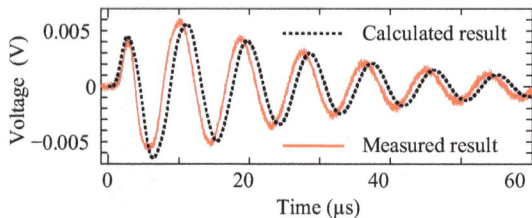

(a)

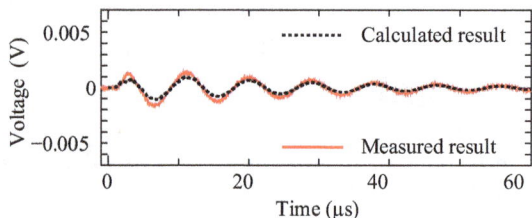

(b)

Figure 11.11   Calculated and measured induced voltages (with capacitors) [30].
(a) Unshielded cable and (b) shielded cable.

analysis space with the soil. In the case of the unshielded control cable, the ground impedance was given using the following expressions [51]:

$$Z_g = \frac{j\omega\mu_0}{2\pi} \ln\left(\frac{1 + \gamma_g h_c}{\gamma_g h_c}\right),$$ (11.7)

$$\gamma_g = \sqrt{j\omega\mu_0\left(\sigma_g + j\omega\varepsilon_0\varepsilon_{rg}\right)},$$ (11.8)

where h_c, ε_0, μ_0, and ω are, respectively, the height of a conductor, the permittivity and permeability in a vacuum, and the angular frequency. σ_g, and ε_{rg} are the conductivity and relative permittivity of soil, respectively. h_c was treated approximately as the height of the center axis of the wire pair. Similarly to the shielded control cable, the p.u.l. inductance and capacitance matrices of the unshielded control cable were obtained through numerical simulations as mentioned in Section 11.3.2.

The aforementioned 3-D FDTD analysis space was modeled using 566 × 912 × 307 non-uniform cells (0.05 to 2 m), whereas the time step was set to 48.1 ps.

11.4.2   Calculated and measured induced voltages

Figures 11.11 and 11.12 show the calculated and measured voltages induced in the secondary circuit at the protection relay with and without the surge-absorbing capacitor for both unshielded and shielded control cables, respectively. In these

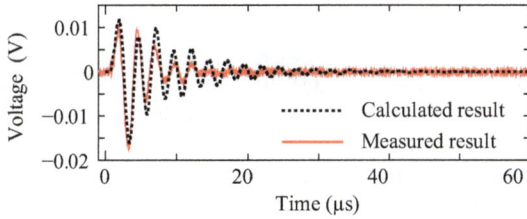

(a)

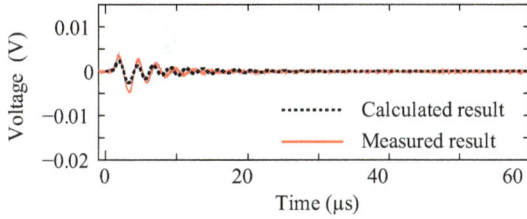

(b)

Figure 11.12   Calculated and measured induced voltages (without capacitors).
(a) Unshielded cable and (b) shielded cable.

results, the peak value of the injected lightning impulse current was normalized to
1 A. As shown in Figures 11.11 and 11.12, the induced voltages are well suppressed
using the surge-absorbing capacitor and the shielded control cable. The calculated
voltages induced on the unshielded control cable with and without the surge-
absorbing capacitor are in good agreement with the measured results. For the
shielded control cable, the differences between the measured and calculated
induced voltages are larger presumably owing to modeling errors of the surface
transfer impedance caused by the deformation of the tape-type shield during
installation, but the calculated waveforms of the induced voltages agree well with
the measured results. These results validate the applicability of the 3-D FDTD
method to the electromagnetic transient analysis of secondary circuits of a substation for lightning impulse currents entering grounding structures. This approach is
very effective for assessing the protection of secondary circuits from lightning-
induced electromagnetic disturbances through the installation of shielded control
cables and surge-absorbing capacitors.

In the above-described measurement and calculation, the current injection wire
was placed parallel to the surface of the soil, and the waveform of the sheath
current on the control cable was dependent mainly on the transient response of the
grounding grid, that is, the effect of the GPRs. On the other hand, in case of
lightning strikes to substations, the sheath current is induced on the metallic sheath
of the control cable also owing to the influence of the lightning electromagnetic
pulse (LEMP) of a return stroke current, and the influence of the LEMP should be
studied in future works.

11.5 Conclusion

The FDTD method has become a very useful tool for analyzing electromagnetic transient phenomena in 3-D structures and grounding systems, thanks to the development of simulation techniques and the progress of high-performance computing, nowadays. Here, using the technique for representing shielded control cables in the 3-D FDTD method, which is commonly employed to protect secondary circuits in substations from electromagnetic disturbances, we calculated voltages induced on a shielded control cable placed over grounding structures when a lightning impulse current flew into the grounding structure, and the FDTD simulations were validated through the comparison with the measured results. These results confirm the applicability of the 3-D FDTD method to the lightning transient analysis of secondary circuits in substations for evaluating the effectiveness of countermeasures based on shielded control cables and surge-absorbing capacitors installed at protection relays.

References

[1] CIGRE WG C4.208, "EMC within power plants and substations," *CIGRE Technical Brochure*, no. 535, 2013.

[2] M. M. Rao, M. J. Thomas, and B. P. Singh, "Transients induced on control cables and secondary circuit of instrument transformers in a GIS during switching operations," *IEEE Trans. Power Del.*, vol. 22, no. 3, pp. 1505–1513, 2007.

[3] D. E. Thomas, C. M. Wiggins, T. M. Salas, F. S. Nickel, and S. E. Wright, "Induced transients in substation cables: measurements and models," *IEEE Trans. Power Del.*, vol. 9, no. 4, pp. 1861–1868, 1994.

[4] H.-T. Wu, X. Cui, X.-F. Liu, *et al.*, "Characteristics of electromagnetic disturbance for intelligent component due to switching operations via a 1100 kV AC GIS test circuit," *IEEE Trans. Power Del.*, vol. 32, no. 5, pp. 2228–2237, 2017.

[5] R. Hatano, T. Ueda, K. Nojima, and H. Motoyama, "Surge voltages induced in secondary circuits of 275 kV full GIS," *IEEJ Trans. Power Energy*, vol. 123, no. 11, pp. 1313–1318, 2003.

[6] A. M. Miri and A. Stojković, "Transient electromagnetic phenomena in the secondary circuits of voltage- and current transformers in GIS (measurements and calculations)," *IEEE Trans. Power Del.*, vol. 16, no. 4, pp. 571–575, 2001.

[7] L. Liu, X. Cui, and L. Qi, "Simulation of electromagnetic transients of the bus bar in substation by the time-domain finite-element method," *IEEE Trans. Electromagn. Compat.*, vol. 51, no. 4, pp. 1017–1025, 2009.

[8] B. Filipović-Grčić, I. Uglešić, V. Milardić, and D. Filipović-Grčić, "Analysis of electromagnetic transients in secondary circuits due to disconnector

switching in 400 kV air-insulated substation," *Elect. Power Syst. Res.*, vol. 115, pp. 11–17, 2014.

[9]   H. Ke, W.-J. Lee, M.-S. Chen, J.-P. Liu, and J. S. Yang, "Grounding techniques and induced surge voltage on the control signal cables," *IEEE Trans. Ind. Appl.*, vol. 34, no. 4, pp. 663–668, 1998.

[10] H. Wu, C. Jiao, and X. Cui, "Study on coupling of very fast transients to secondary cable via a test platform," *IEEE Trans. Electromagn. Compact.*, vol. 60, no. 5, pp. 1366–1375, 2018.

[11] Electrotechnical Research Association, "Technologies of countermeasures against surges on protection relays and control systems," *ETRA Report*, vol. 57, no. 3, 2002 (in Japanese).

[12] A. Ametani, H. Motoyama, K. Ohkawara, H. Yamakawa, and N. Suga, "Electromagnetic disturbances of control circuits in power stations and substations experienced in Japan," *IET, Generation, Transmission & Distribution*, vol. 3, no. 9, pp. 801–815, 2009.

[13] K. S. Yee, "Numerical solution of initial boundary value problems involving Maxwell's equations in isotropic media," *IEEE Trans. Antennas Propag.*, vol. AP-14, no. 3, pp. 302–307, 1966.

[14] R. F. Harrington, *Field Computation by Moment Methods,* Macmillan, New York, 1968.

[15] J.-M. Jin and D. J. Riley, *Finite Element Analysis of Antenna and Arrays,* Wiley, Hoboken, NJ, 2008.

[16] CIGRE Working Group C4.501, "Guideline for numerical electromagnetic analysis method and its application to surge phenomena," *CIGRE Technical Brochure*, no. 543, 2013.

[17] A. Tatematsu and T. Noda, "Three-dimensional FDTD calculation of lightning-induced voltages on a multiphase distribution line with the lightning arresters and an overhead shielding wire," *IEEE Trans. Electromagn. Compact.*, vol. 56, no. 1, pp. 159–167, 2014.

[18] M. Namdari, M. K.-Farsani, R. Moini, and S. H. H. Sadeghi, "An efficient parallel 3-D FDTD method for calculating lightning-induced disturbances on overhead lines in the presence of surge arresters," *IEEE Trans. Electromagn. Compact.*, vol. 57, no. 6, pp. 1593–1600, 2015.

[19] A. Tatematsu and T. Ueda, "FDTD-based lightning surge simulation of an HV air-insulated substation with back-flashover phenomena," *IEEE Trans. Electromagn. Compact.*, vol. 58, no. 5, pp. 1549–1560, 2016.

[20] H. Motoyama, "Experimental study and analysis of breakdown characteristics of long air gaps with short tail lightning impulse," *IEEE Trans. Power Del.*, vol. 11, no. 2, pp. 972–979, 1996.

[21] CIGRE Working Group C4.37, "Electromagnetic computation methods for lightning surge studies with emphasis on the FDTD method," *CIGRE Technical Brochure*, no. 785, 2019.

[22] A. Tatematsu, Y. Baba, M. Ishii, S. Okabe, T. Ueda, and N. Itamoto, "Development of surge simulation techniques based on the finite difference

time domain method and its application to surge analysis," *CIGRE Session 2016*, no. C4-302, 2016.

[23]   Y. Baba and V. A. Rakov, "Applications of the FDTD method to lightning electromagnetic pulse and surge simulations," *IEEE Trans. Electromagn. Compat.*, vol. 56, no. 6, pp.1506–1521, 2014.

[24]   A. Tatematsu, S. Moriguchi, and T. Ueda, "Switching surge analysis of an EHV air-insulated substation using the 3-D FDTD method," *IEEE Trans. Power Del.*, vol. 33, no. 5, pp. 2324–2334, 2018.

[25]   A. Tatematsu, S. Terakuchi, T. Yanagi, T. Kamibayashi, and R. Mori, "Lightning current simulation of 66-kV substation with power cables using the three-dimensional FDTD method," *IEEE Trans. Electromagn. Compat.*, vol. 63, no. 3, pp. 819–829, 2021.

[26]   J. Takami, T. Tsuboi, K. Yamamoto, S. Okabe, Y. Baba, and A. Ametani, "Lightning surge into a substation at a back-flashover and review of lightning protective level through the FDTD simulation," *IEEE Trans. Dielectr. Electr. Insul.*, vol. 21, no. 3, pp. 1044–1052, 2014.

[27]   J. Takami, T. Tsuboi, K. Yamamoto, S. Okabe, and Y. Baba, "FDTD simulation considering an AC operating voltage for air-insulation substation in terms of lightning protective level," *IEEE Trans. Dielectr. Electr. Insul.*, vol. 22, no. 2, pp. 806–814, 2015.

[28]   A. Tatematsu, K. Yamazaki, K. Miyajima, and H. Motoyama, "A study on induced voltages on an aerial wire due to a current flowing through a grounding grid," *IEEJ Trans. Power Energy*, vol. 129, no. 10, pp. 1245–1251, 2009.

[29]   A. Tatematsu, F. Rachidi, and M. Rubinstein, "Three-dimensional FDTD-based simulation of induced surges in secondary circuits owing to primary-circuit surges in substations," *IEEE Trans. Electromagn. Compat.*, vol. 63, no. 4, pp. 1078–1089, 2021.

[30]   A. Tatematsu and A. Yamanaka, "Three-dimensional FDTD-based simulation of lightning-induced surges in secondary circuits with shielded control cables over grounding grids in substations," *IEEE Trans. Electromagn. Compat.*, vol. 65, no. 2, pp. 528–538, 2023.

[31]   A. Tatematsu, "A technique for representing lossy thin wires and coaxial cables for FDTD-based surge simulations," *IEEE Trans. Electromagn. Compat.*, vol. 60, no. 3, pp. 705–715, 2018.

[32]   Y. Du, B. Li, and M. Chen, "The extended thin wire model of lossy round wire structures for FDTD simulations," *IEEE Trans. Power Del.*, vol. 32, no. 6, pp. 2472–2480, 2017.

[33]   M. Feliziani and F. Maradei, "Full-wave analysis of shielded cable configurations by the FDTD method," *IEEE Trans. Magn.*, vol. 38, no. 2, pp. 761–764, 2002.

[34]   B. Li, Y. Ding, Y. Du, and M. Chen, "Stable thin-wire model of buried pipe-type power distribution cables for 3D FDTD transient simulation," *IET Gen. Trans. Dist.*, vol. 14, no. 25, pp. 6168–6178, 2020.

[35] B. Li, Y. Du, M. Chen, and Z. Li, "A 3-D FDTD thin-wire model of single-core coaxial cables with multiple conductive layers," *IEEE Trans. Electromagn. Compat.*, vol. 63, no. 3, pp. 762–771, 2021.

[36] A. Tatematsu, "Lightning surge analysis of a transmission line tower with an XLPE power cable and metallic cleats using the FDTD method," *IEEE Trans. Electromagn. Compat.*, vol. 62, no. 5, pp. 1796–1806, 2020.

[37] E. F. Vance, *Coupling to Shielded Cables,* Wiley, New York, 1978.

[38] S. Celozzi and M. Feliziani, "FDTD analysis of the interaction between a transient EM field and a lossy shielded cable," in *Proc. 10th Int. Zurich Symp. Electromagn. Compat.*, Zurich, Switzerland, March 9–11, pp. 493–498, 1993.

[39] T. Noda, "Identification of a multiphase network equivalent for electromagnetic transient calculations using partitioned frequency response," *IEEE Trans. Power Del.*, vol. 20, no. 2, pp. 1134–1142, 2005.

[40] Z. P. Liao, H. L. Wong, B. P. Yang, and Y. F. Yuan, "A transmitting boundary for transient wave analysis," *Sci. Sin. A*, vol. 27, no. 10, pp. 1063–1076, 1984.

[41] T. Noda and S. Yokoyama, "Thin wire representation in finite difference time domain surge simulation," *IEEE Trans. Power Del.*, vol. 17, no. 3, pp. 840–847, 2002.

[42] Y. Baba, N. Nagaoka, and A. Ametani, "Modeling of thin wires in a lossy medium for FDTD simulations," *IEEE Trans. Electromagn. Compat.*, vol. 47, no. 1, pp. 54–60, 2005.

[43] M. P.-May, A. Taflove, and J. Baron, "FD-TD modeling of digital signal propagation in 3-D circuits with passive and active loads," *IEEE Trans. Microw. Theory Tech.*, vol. 42, no. 8, pp. 1514–1523, 1994.

[44] A. Tatematsu, K. Yamazaki, and H. Matsumoto, "Lightning surge analysis of a microwave relay station using the FDTD method," *IEEE Trans. Electromagn. Compat.*, vol. 57, no. 6, pp. 1616–1626, 2015.

[45] P. de Arizon and H. W. Dommel, "Computation of cable impedances based on subdivision of conductors," *IEEE Trans. Power Del.*, vol. 2, no. 1, pp. 21–27, 1987.

[46] P. Oeding and K. Feser, "Geometric mean distances of rectangular conductors," *ETZ-A*, vol. 86, no. 16, pp. 525–533, 1965.

[47] T. Miki and T. Noda, "An improvement of a conductor subdivision method for calculating the series impedance matrix of a transmission line considering the skin and proximity effects," *IEEJ Trans. Power Energy*, vol. 128, no. 1, pp. 254–261, 2008 (in Japanese).

[48] A. Tatematsu, F. Rachidi, and M. Rubinstein, "On the representation of thin wires inside lossy dielectric material for FDTD-based LEMP simulations," *IEEJ Trans. Electr. Electro. Eng.*, vol. 14, no. 9, pp. 1314–1322, 2019.

[49] A. Tatematsu, F. Rachidi, and M. Rubinstein, "A technique for calculating voltages induced on twisted-wire pairs using the FDTD method," *IEEE Trans. Electromagn. Compat.*, vol. 59, no. 1, pp. 301–304, 2017.

[50]   A. K. Agrawal, H. J. Price, and S. Gurbaxani, "Transient response of mul-
        ticonductor transmission lines excited by a nonuniform electromagnetic
        field," *IEEE Trans. Electromagn. Compat.*, vol. EMC-22, no. 2, pp. 119–129,
        1980.
[51]   F. M. Tesche, M. V. Ianoz, and T. Karlsson, *EMC Analysis Methods and
        Computational Models,* Wiley, New York, 1997.

Chapter 12

An introduction to a resilience-based approach to transient high-power electromagnetic disturbance mitigation

Richard Hoad[1]

12.1 Introduction

This chapter introduces the concept of a resilience-based approach rather than a protection-lead approach to transient High Power Electromagnetic (HPEM) disturbance mitigation. In this chapter, it is shown that a resilience-based approach is much more intuitive and is how the risk from the majority of common hazards such as fire hazards for instance, are mitigated. The resilience-based approach and a framework for the management of HPEM resilience [1] have been developed for an update and conversion to an International Standard (IS) courtesy of the International Electrotechnical Commission (IEC) to IEC 61000-5-6 [2].

12.2 The protection-dominated approach

The earliest standards for High-altitude Electromagnetic Pulse (HEMP) protection of facilities were developed for military use [3,4]. This is at least partially because the HEMP threat was first acknowledged by the military sector which required a zero-disruption or "work-through" solution for facilities that could be exposed to the HEMP threat. Military HEMP protection standards generally require a high degree of protection, mainly due to the requirement to work-through HEMP events, and also due to the inclusion of a significant protection margin so that degradation in the protection performance does not impact the ability of the facility to work-through.

However, military electronic equipment tends to operate in a well-bounded-networked manner and is resilient by design. Military facilities that are HEMP-protected tend to be continuously staffed with trained professionals who may have dedicated responsibilities for the maintenance of HEMP protection. The protection-led approach is known to be very effective as long as the "as-built" protection performance is maintained and continuously assured.

[1]QinetiQ Ltd., Cody Technology Park, Farnborough, Hants, UK

As long as the protection level (with margin) is high enough to reduce the threat to a level that cannot disturb the equipment within the facility then the approach is entirely valid. Figure 12.1 attempts to demonstrate this condition. The yellow line represents the mission or operation of the facility. When an HPEM disturbance impinges on the facility there is no negative mission impact.

The cost of protection and the effort required to continually maintain and verify that the protection level is adequate can be significant for this condition. For this protection-dominated or work-through approach, a response and recovery plan may not have been produced since there is no expectation of protection degradation or failure. However, if the protection has failed or is degraded, which could be due to many factors, then this could lead to an exaggerated and likely undesirable recovery time for the mission. This condition is shown in Figure 12.2. This time the

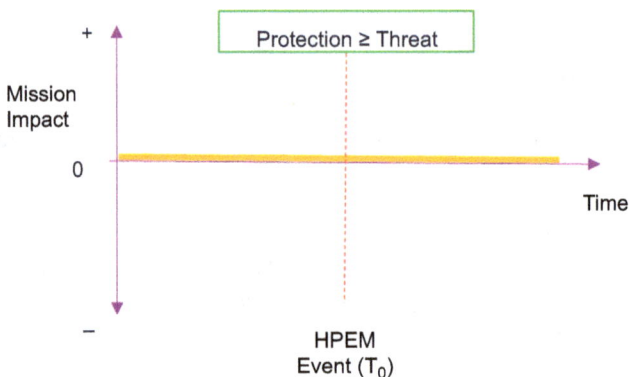

Figure 12.1   Mission impact for the condition when the protection level including margin is equal to or exceeds the HPEM threat

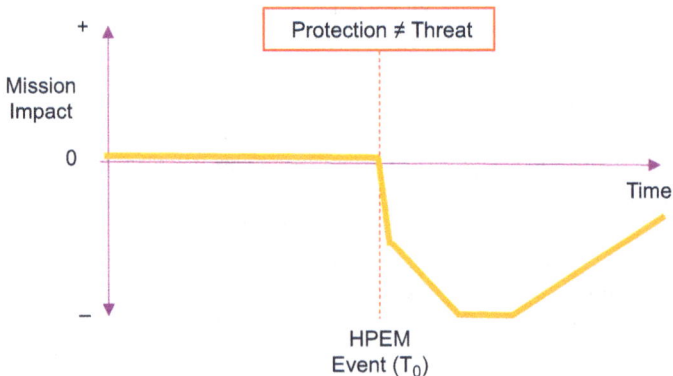

Figure 12.2   Mission impact for the condition when the protection level is not sufficient to mitigate the HPEM threat

HPEM disturbance event has a profound negative impact on the mission which could last for a significant time.

12.2.1   The disadvantages of a protection-dominated approach

The protection-dominated, work-through, approach, described above can be cost-intensive, inefficient, and very difficult to apply for modern applications and facilities. On the contrary, the proposed resilience-based approach focuses on "resilience" and has a wide range of application scenarios compared to the protection-dominated approach.

Ultimately a protection or resilience scheme has to be affordable. Affordability is subjective but is a function of the cost/impact of failure of the system (risk) during the system's lifetime. This must be contrasted against the likelihood that the risk will occur which in turn is a function of the occurrence of the threat. HPEM disturbances have a very low likelihood of occurrence but the potential impact on an infrastructure system could be extremely severe [5].

The cost of EM protection not only includes the up-front or capitalization costs, but also the whole-life costs of the protection scheme which includes the cost of continued assurance and the cost of remedial repairs to a damaged protection scheme.

A modern facility undergoes regular changes or upgrades. For example, many Information Communication Technology (ICT) suppliers recommend a review and refresh cycle that modernizes ICT infrastructure every 2 to 4 years. Other aspects such as new energy efficiency initiatives can also drive the requirement for regular upgrades. This means that infrastructure owners/operators are in an almost constant cycle of upgrades to meet the latest requirements. Unless great care is taken during an upgrade which will inevitably require new penetrations through the protected boundary, a protection scheme can very easily be compromised.

Due to the factors discussed above, "work-through" protection schemes are not widely adopted by infrastructure providers for HPEM disturbances. Ultimately this is because of uncertainty over the risk and uncertainty of the likelihood of occurrence of the threat and it means that protection for HPEM is only applied in very limited cases. The "as-built" and "whole life" costs of protection [6] and the difficulty managing upgrades discourage many from applying for protection at all.

12.3   A resilience-based approach

The definition of the term Resilience (Electromagnetic), used in [2], is: "features of a system that enable the system to be prepared for, to withstand, to respond to and to recover from a transient electromagnetic disturbance(s) in a timely and efficient manner". Note that this definition of resilience includes the requirement for protection (withstand) but other attributes are added. A key change here is a shift in emphasis implied in the protection-led approach from "shall continue to work-through," to a "shall be capable of timely recovery" emphasis in the resilience-based approach. For the resilience-based approach, there is an implied acceptance that the

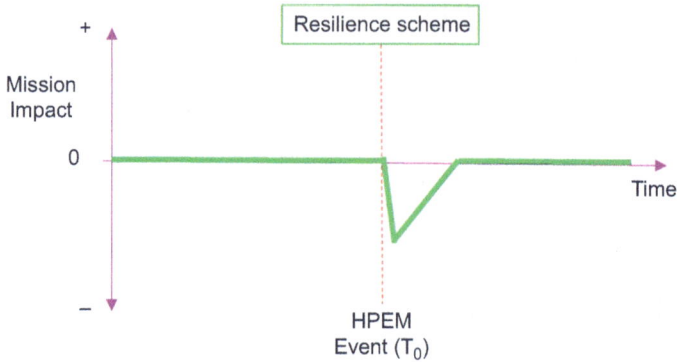

Figure 12.3   Mission impact for the condition when a resilience-based approach is used. Note that the recovery time is faster than that in Figure 12.2.

mission or function of a system or facility may be affected or disrupted and therefore that prompt restoration and recovery are likely to be required.

Ultimately, the vast majority of systems or facilities in a modern infrastructure are not required to remain functional or work-through an HPEM disturbance but they must be protected sufficiently so that they can be restored in a timely and efficient manner after the event has subsided.

The condition when a resilience-based approach is successfully applied is shown in Figure 12.3. This time the HPEM disturbance event still has a negative impact on the mission but the response and recovery time is manageable since the facility is prepared to recover from the HPEM event.

12.3.1   A model for HPEM resilience

A simple model that describes the necessary attributes or functions that can be used to develop a framework-based approach for a resilience-based approach is shown in Figure 12.4. This model is derived from existing, authoritative, and peer-reviewed principles and practices developed for cyber threats described in the National Institute of Standards and Technology (NIST) Cyber Security Framework [7].

The resilience-based approach is intuitive and is used for many other types of common threats. Consider for example the risk of a fire to a populated building, e.g., a school, hotel, or office. Full fire protection of the building or even a zone within a building may be technically feasible but may severely limit accessibility and may be unaffordable. The residual risk that a fire will damage the building or worse claim lives cannot be fully mitigated by protection alone, though protection such as fire-retardant materials, fire doors, and fire suppression systems are likely to be implemented. Because of this residual risk and because the people at risk do not need to work-through the fire incident, fire and smoke detection is implemented and there will be occasional tests of the system and evacuation drills, essentially training and rehearsal. The fire protection solutions merely slow the spread of the fire to buy time for the emergency services to intervene.

Figure 12.4   Resilience model functions

However, the resilience-based approach is not applicable to all systems as there is a clear dependency on system criticality or system availability requirements, or both. Table 12.1 indicates where the resilience-based has the most applicability against system availability/recovery time.

12.3.2   Brief description of the resilience model functions

The five functions included in the framework core are: Identify, Protect, Detect, Respond, and Recover. These functions are briefly described below.

12.3.2.1   The Identify function

The Identify function promotes an organizational understanding of managing risk to systems, people, assets, data, and capabilities. Understanding the business context, the resources that support critical functions, and the related risks enables an organization to focus and prioritize its efforts, in accordance with its risk management strategy and business needs.

12.3.2.2   The Protect function

The Protect function outlines appropriate safeguards to ensure the delivery of critical infrastructure services. The Protect Function supports the ability to limit or contain the impact of a potential event.

12.3.2.3   The Detect function

The Detect function defines the appropriate activities to identify the occurrence of a cybersecurity event. The Detect function enables the timely discovery of events. It also enables the call-to-action of first responders—those individuals who attend to the response and recovery activities. HPEM detection is discussed in another chapter of this book.

Table 12.1   Guide to the appropriate application of the resilience-based approach

	Acceptable response and recovery time				
	Weeks	**Days**	**Hours**	**Minutes**	**Seconds/ work-through**
Suggested approach to EM disturbance mitigation	Resilience-based approach	Resilience-based approach	Resilience-based approach	Resilience-based approach has some benefits but protection becomes an important consideration	Protection dominated with margin, redundancy and preventative maintenance scheme
Reliance on protection technology	Very low	Low—protection from damage unless spares are readily available	Moderate—protection from damage	High—protection from damage; some disruption can be tolerated	Very high—protection from damage and disruption with a protection margin
Applicability of detection	Provides benefits in directing first responders	Provides benefits in directing first responders	Provides benefits in directing first responders	Provides benefits in directing first responders	Provides limited benefits[a]
Reliance on first responders for recovery	Low	Moderate—first responders are likely to require their own mitigations	High—first responders are likely to require their own mitigations	High—first responders will require their own mitigations	Low

[a]The resilience-based approach can have benefits to systems in the right-hand column as the approach will enable response and recovery actions in the event that the protection has failed or is unknowingly compromised.

12.3.2.4   The Respond function

The Respond function includes appropriate activities to take action, including those of first responders, regarding a detected incident. The Respond function supports the ability to contain the impact of a potential incident.

12.3.2.5   The Recover function

The Recover function identifies appropriate activities to maintain plans for resilience and to restore any capabilities or services that were impaired due to an incident. The Recover function supports timely recovery to normal operations to reduce the impact of an incident.

The five functions describe activities that are further subdivided into categories and sub-categories. These are summarized in an Excel Spreadsheet [8]. These collectively drive outcomes, which in turn lead to successful management of resilience, in the cyber context.

The NIST cyber security framework is highly adaptable and flexible and has been adapted as the basis for an HPEM resilience approach, which is fully described in the latest published version of IEC 61000-5-6 Ed 1.0 International Standard [2].

12.4 Summary

The resilience-based approach is consistent with the practices employed by many modern critical facilities where, for business continuity reasons, there is very often a clear business continuity plan and/or a response and recovery strategy.

A key difference in the resilience-based approach from the protection-dominated approach is that the resilience-based approach requires consideration of the recovery time of the mission or system. Due to the lack of consideration for the response and recovery in the protection-dominated approach, the period for response and recovery of the mission or system could be significant. However, the period for response and recovery of the system with the resilience-based approach is manageable since the mission or system is prepared to recover from the HPEM event.

The resilience-based approach and a framework for the management of resilience to HPEM threats have been developed for an update to IEC 61000-5-6 Ed. 1.0 which was published as an International Standard in April 2024.

References

[1] R. Hoad and B. Petit, 'A Resilience-based Approach to HPEM Threat Mitigation', *The GlobalEM Conference 2022*, Abu Dhabi, United Arab Emirates, 2022.

[2] IEC 61000-5-6 Ed. 1.0 (2024-04-05): Electromagnetic Compatibility (EMC) – Part 5-6: Installation and Mitigation Guidelines – Mitigation of External EM Influences.

[3] R. Hoad and W. A. Radasky, 'Progress in High-Altitude Electromagnetic Pulse (HEMP) Standardization.', *IEEE Transactions on Electromagnetic Compatibility*, Vol. 55, No. 3, pp. 532–538, 2013.

[4] Mil-Std-188-125-1, 'High-Altitude Electromagnetic Pulse (HEMP) Protection for Ground-Based C4I Facilities Performing Critical, Time-Urgent Missions,' *Part 1, Fixed Facilities*, 2005.

[5] Report of the Commission to Assess the Threat to the United States from Electromagnetic Pulse (EMP) Attack, Critical National Infrastructures, April 2008 (download at www.empcommission.org).

[6]   Public Submission to the US National Reliability Council, 'Comments of the Foundation for Resilient Societies, Inc. on Mitigation Strategies for Beyond-Design-Basis Events', NRC Docket No. NRC-2014-0240, 80 FR 70609.

[7]   NIST, 'Cyber Security Framework', [Online]. Available: https://doi.org/10.6028/NIST.CSWP.04162018 [Accessed 2022].

[8]   NIST, 'NIST Cyber Security Framework – Core Elements', [Online]. Available: https://www.nist.gov/document/2018-04-16frameworkv11core1xlsx.

Chapter 13

Susceptibility of switch-mode power supplies due to conducted pulse from HEMP

Laurine Curos[1], Guillaume Mejecaze[1], Tristan Dubois[2], Jean-Michel Vinassa[2] and Frédéric Puybaret[1]

In the past years, the threat of electromagnetic pulses caused by a high-altitude nuclear explosion is still an actual concern in the field of security and safety [1]. In fact, E1 High-altitude ElectroMagnetic Pulse (E1-HEMP) couples efficiently on aerial lines of the electricity network allowing it to supply houses and factories [2]. The conducted parasitic currents, of hundreds of amperes, and voltages, of several kilovolts, induced by the coupling of an E1 HEMP on an electrical network involve disturbances and even destruction of equipment's power supplies plugged into the grid [3,4].

Nowadays, the majority of critical infrastructures, such as electricity distribution, telecommunications, health facilities, etc., are composed of many electronic devices necessary for functional operation and to carry out their critical functions. Therefore, these infrastructures are dependent on continuous electricity supply. In a suitable attack scenario of HEMP, the increase in electronics implies an increase in the vulnerability of critical infrastructure. Electronic structures are often not certified beyond their frequency band or voltages/currents ranges as high as in an HEMP scenario. As can be seen in Figure 13.1, the frequency bandwidth of the HEMP covers the majority of listed disturbance sources as lightning or electrostatic discharges (ESD). Due to the wide frequency bandwidth, from quasi-continuous to a few tens or even hundreds of megahertz, the HEMP can impact a large amount of equipment, small and large equipment. Tests on systems of larger dimensions, such as buildings, have shown the importance of the conducted mode. Indeed, even if few publications are present in the literature, the conducted mode appears to be the most destructive mode [5].

In this domain, most of the studies available in the literature deal with the effects of EMP on large systems or integrated circuits such as microcontrollers and logic circuits [6,7]. In most cases, if we look deeply inside an electrical consumer or industrial equipment structure plugged into the grid, the first element that will be

[1]CEA DAM, CEA-Gramat, France
[2]Université de Bordeaux, CNRS, Bordeaux INP, IMS UMR 5218, France

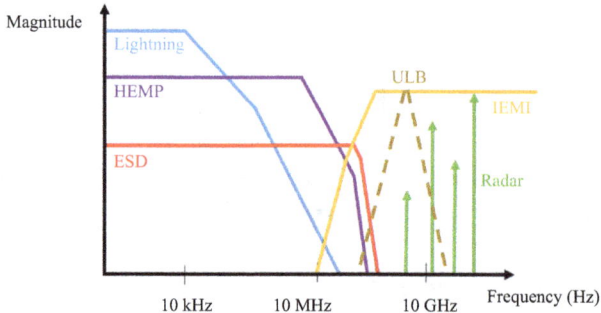

Figure 13.1   Spectral range of some electromagnetic interferences

struck by the current disturbance is the Switch-Mode Power Supply (SMPS) [8]. Due to the system's complexity, most of the time, only a few studies propose an analysis of the destruction effects at the component level [9,10]. However, in a HEMP scenario, the precise understanding of the mechanisms leading to the equipment destruction and especially to the power supply destruction is still insufficient to be correctly anticipated.

In this context, the chapter describes susceptibility research and the different steps used to predict SMPS destruction due to the current pulse induced by the coupling between HEMP and network distribution wires.

13.1   Approach in susceptibility studies

An approach in susceptibility study could be to consider a system, a power supply in our case, like "a black box". A disturbance is injected at the system input, and it is observed whether a failure occurs or no effect is seen. It is a go/no-go test. In this case, explaining the failure of the studied global system and reproducing it in numerical simulation could be difficult.

How to predict numerically the electromagnetic interference effects on electronic power supplies? In fact, understanding precisely the effect on the studied system is the key. An approach is developed in this way and presented in the diagram in Figure 13.2.

The first step of our approach is to analyze the commercial Switch-Mode Power Supply (SMPS) topologies. In fact, reducing the number of test cases in the first instance is important to focus on understanding the failure mechanism. A study of a wide range of commercial SMPS topologies and components has been performed to design a representative power supply for testing. This study has permitted to study of a known SMPS topology and develop the approach on an understood equipment. Generalization to an entire SMPS power class will occur in further work. Then, experimentations are performed on the designed SMPS. The current pulse induced by the coupling between HEMP and the electrical network is injected at the SMPS input thanks to an injection generator (detailed in Section 13.2.2) until destruction.

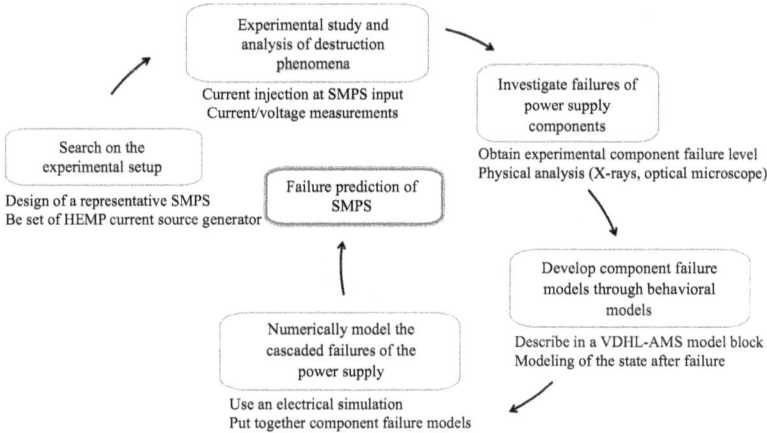

Figure 13.2 Approach diagram applied to the SMPS destruction study

For this second step, current and voltage measurements at different specific nodes of the SMPS during the pulse injection permit us to understand and propose chronological failure events. After that, for the most susceptible components in power supplies, analysis of the datasheet maximum ratings and analysis of failed components (X-rays, optical microscope) are realized in order to develop failure models. It is the third step presented in the diagram (Figure 13.2).

The penultimate step consists in developing component failure models based on the previous step. Using the experimental failure levels, a behavioral model of each component is developed.

The approach concludes by associating the developed failure models of each component. Consequently, we could be able to simulate numerically the component failure and therefore, predict the SMPS destruction effects induced by high-level current pulse injection on power supplies.

13.2 Experimental setup

13.2.1 Studied power supply

Power supplies connected to the mains can be of various topologies [11] depending on the output power. The application note [12] resumes the power supply topologies associated with the output power range. The graphic in Figure 13.3 brings out two output power ranges for the main operated AC/DC SMPS: from 0 W to 150 W and from 150 W to 1 kW. Thus, for power usage less than 100 W, referred to as "general public" power supplies, flyback topology is mainly used. This result has been confirmed by G. Mejecaze [13] after a SMPS panel analysis. Nowadays with the power reduction trend, more and more SMPS based on flyback topology are used.

In this way, after a components study on SMPS of output power range less than 100 W, a representative power supply in terms of technology and sizing of

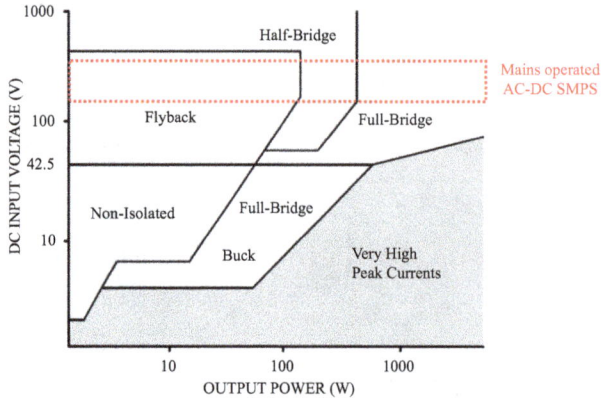

Figure 13.3   *Topologies of power supplies according to their output power [11]*

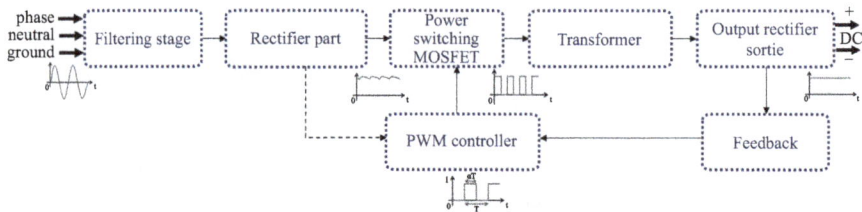

Figure 13.4   *Diagram of the designed SMPS*

electronic components is designed. Details about commercial SMPS panel analysis are given in [10]. The generic power supply considered here is an AC/DC switching power supply of flyback topology, with a 19 V/3 A output, being part of a power range less than 100 W. Figure 13.4 shows the block diagram composed of all the functions of the used flyback power supply. At the input, the rectifier stage composed of a bridge rectifier and a smoothing capacitor, placed after the EMC filter, rectifies the sinusoidal mains voltage to have a quasi-continuous voltage. The PWM switching controller, also known as Pulse Width Modulation, ensures the switching control of the Mosfet transistor at a frequency of 65 kHz and adapts the duty cycle according to the output feedback loop, known as Feedback. Galvanic isolation is ensured by the high-frequency transformer.

In order to perform several destructive tests, this power supply have been designed and many samples were manufactured.

13.2.2   Injection source

A current injection source, presented in Figure 13.5, has been designed to generate a high current pulse. This pulse shape is determined by considering the radiated HEMP standard IEC (High Amplitude Electromagnetic Pulse – IEC 61000-2-9 [14]).

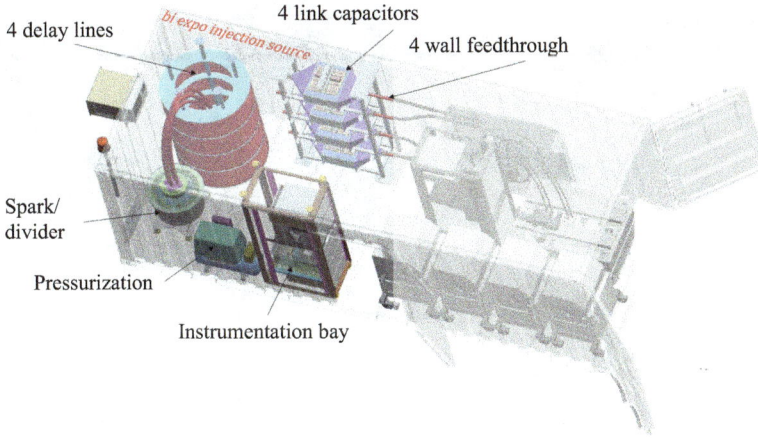

Figure 13.5   Current injection source

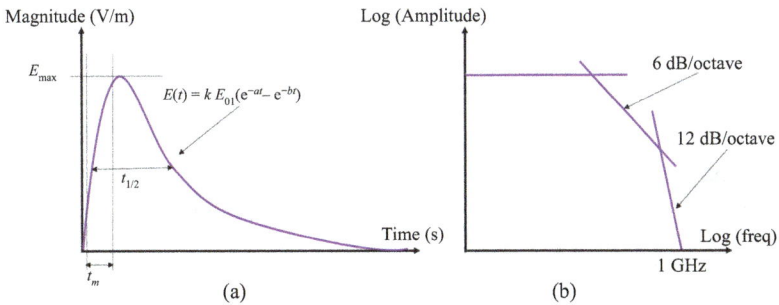

Figure 13.6   Injected bi-exponential shape in the time domain (a) and in the frequency domain (b)

From the incident field described in this standard, the relevant current shape conducted through equipment connected to the mains is calculated using our own coupling calculation on electric wires in a 3D electromagnetic simulation proprietary software. The generated signal is a bi-exponential type configurable in rise time, mid-height time, and level to analyze their influence on the susceptibility. An example of this signal is presented in Figure 13.6.

More details about the current injection source (Figure 13.5) are done by Mejecaze in [13]. This current source is associated with an electrical generation-transformation-distribution network, representative of the conventional distribution network, presented in Figure 13.7. This global platform, completely autonomous, is composed of a 160 kVA three-phase generator, a 400 V/20 kV step-up transformer, a 20 kV disconnector, and a 20 kV/400 V step-down transformer. Next, a line impedance stabilization network (LISN) permits to control the impedance upstream of the equipment under test (EUT).

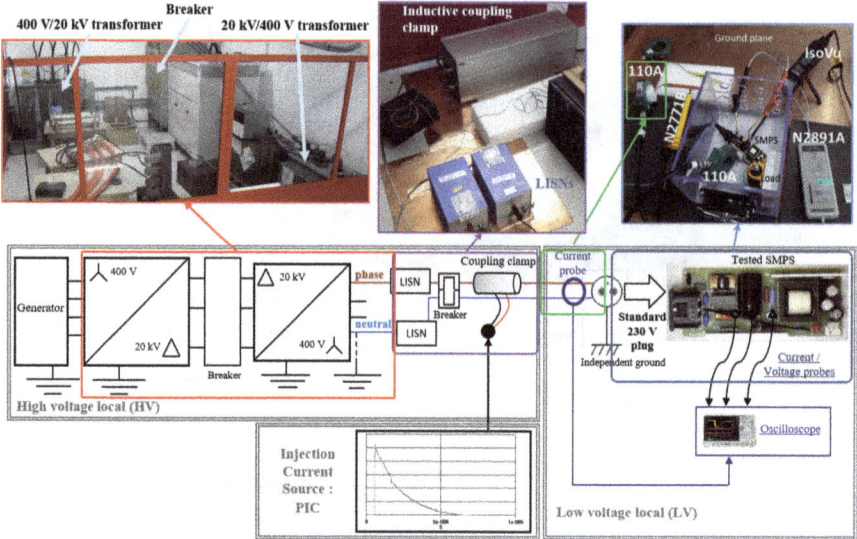

Figure 13.7   Current injection source

The coupling clamp, placed downstream of the LISNs, allows the injection of the disturbance generated by the current injection source (Figure 13.5) on the EUT: the power supply. Therefore, this platform allows the injection of conducted disturbances, including high-current pulses of several hundred amperes for a few nanoseconds, into electronic equipment.

This source gives the possibility to inject the current at tested system input representative of calculated disturbances induced by the coupling of the NEMP field on long-distance wire links.

13.3   SMPS experimental destructions

In the case of a nuclear explosion at a high altitude, current disturbances can be generated and propagated to the electrical network conductors in buildings. These currents can then disturb or even destroy the power supplies of electronic systems connected to the grid. Destruction tests on power supplies, in the publication by James *et al.* [15], make it possible to identify the most often destroyed components within a power supply: the fuse, the rectifier stage with the diodes bridge, the input filter, and the MOSFET transistor. This trend is also confirmed in the work of Mejecaze *et al.* [16]. Moreover, for power supplies output less than 100 W, the most often destroyed components are the same both for a differential mode infection and a common mode one. The graph in Figure 13.8 shows that the rectifier stage input of the power supply, as well as the integrated circuit PWM controller and the associated power transistor (MOSFET transistor), often fail. The fuse fails

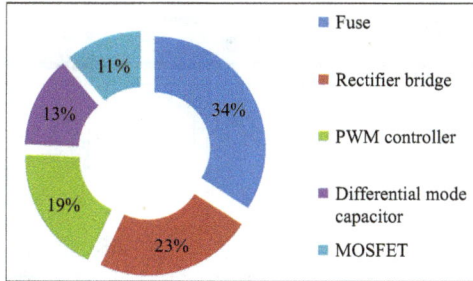

*Figure 13.8   Percentage of the most frequently failed components within the
consumer power supplies <100 W*

in the majority of cases due to the failure in short-circuit of one or more other components.

In order to understand the mechanisms leading to each component failure and so the destruction of the whole SMPS, current and voltage measurements have been specially performed during the destruction (injection of several hundred amperes). Measurements inside SMPS during the "struck" require probes to be able to measure high amplitude signals, reaching several hundreds of amperes and several thousands of volts, with a frequency bandwidth of several tens of megahertz. More details about chosen probes are given in [16]. Susceptibility tests were conducted on a power supply loaded with a 6.3 Ω resistor to be in its nominal operation (57 W: 19 V/3 A) and led to the development of a most likely complete destruction scenario due to a high-level electric pulse injected in differential mode at its input. For this study, injection levels are progressively increased, with 500 V step from the first level (8 kV) set on the injection source (described in Section 13.2.2), until reaching the SMSP destruction threshold. The destruction is observed when the SMPS output voltage is 0 V. After a hundred SMPS destructions, a scenario giving the chronology of the destruction events has been built [17]. The chronology has been determined through these measurements carried out at different specific nodes during the destruction and by analyzing the datasheet maximum ratings and also the destroyed components, using X-rays and an optical microscope. Figure 13.9 shows the sequence of failures presented in [16] when a pulse of several hundred amperes is injected in differential mode at the power supply input. The power supply is destroyed in a few milliseconds after the disturbance injection.

The electromagnetic susceptibility of equipment, such as the power supply subjected to the electrical stress resulting from the coupling of the HEMP, is mainly determined by the electronic components that compose it. In this context, knowledge of the main failure modes of destroyed components in a power supply is necessary.

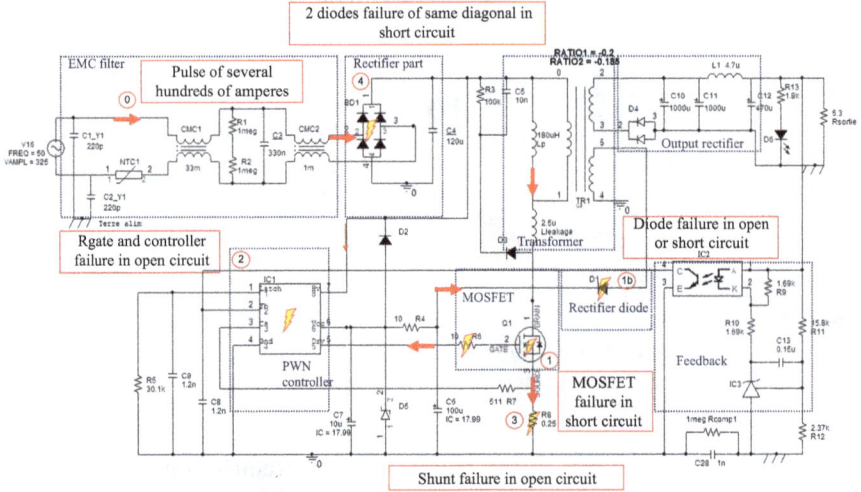

Figure 13.9   Most likely destruction scenario of a power supply by injecting, in differential mode, the result of the coupling of an IEMN-HA on the electrical network

13.4   To the modeling of the SMPS destruction

To predict the SMPS failure level using simulation tools, it is important to study and understand the effects of such a high amplitude current pulse. For this, it is interesting to focus on failure modes to which SMPS components may be subjected. The taken approach aims to answer the following questions. Should we consider modeling a threshold of current, voltage, power, or energy? What is the state of the component once failed? In this way, for each destroyed component in the SMPS, the aim is to determine the failure levels and develop a predictive component failure model. The assembly of each developed model will permit to reproduction of the SMPS cascaded failure scenario.

During tests consisting of injecting several hundred amperes current pulses at power supply input, it has been observed that the disturbance flows through all the SMPS functions causing the destruction of several components. Thus, the manufacturer references of components, identical to those in the power supply, are subjected to electrical pulses. The chronological events of the destruction effects (Figure 13.9) have shown that the MOSFET transistor is the first destroyed element in SMPS. In this section, we focus on this component.

13.4.1   MOSFET failure investigation

During "struck" on power supply, the voltage/current constraints are beyond limitations indicated by the manufacturer. In fact, these constraints can reach a hundred amperes and/or volts with pulse widths ranging from microseconds to milliseconds, as specified in the work in [16]. In this case, the failure of the

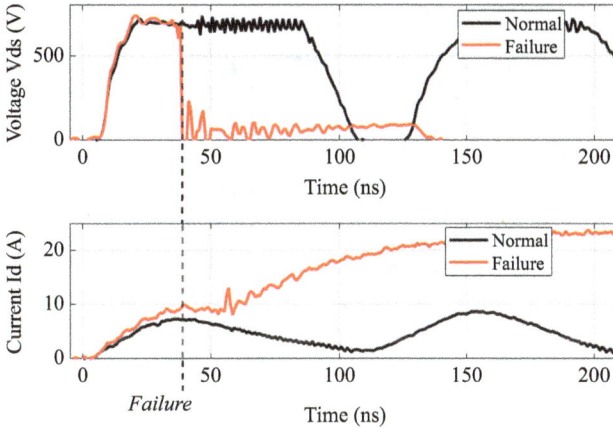

Figure 13.10   MOSFET voltage/current measurements during avalanche tests

MOSFET transistor can be caused by the component avalanche due to an over-voltage between the drain and source pins [18]. This phenomenon generated by a high voltage applied to the drain is known and studied in the bibliography [19]. To produce electrical stress on MOSFET terminals, the injection source, presented in Section 13.2.2, is used to inject various electric pulses. In fact, an increasing voltage signal between the drain and source (gate short-circuited) is injected step by step until failure. Figure 13.10 shows the failure limit due to the avalanche phenomenon [20]. The voltage V_{ds}, between MOSFET's drain and source, clamps to 730 V and a drain current I_d of 7 A flows. At these levels, no failure is observed on the component. By incrementing the generator level, the current flowing in the drain increases, reaching 9 A for a V_{ds} voltage of 730 V. As Figure 13.10 shows, at this moment, the V_{ds} voltage suddenly drops to zero while the current I_d increases: the MOSFET transistor is then failed in short circuit.

Associated with measurements, X-ray pictures and microscopy analyses have been performed on destroyed transistors to observe failures in MOSFET structure and confirm the avalanche phenomenon. The technical datasheet of MOSFET SPA11N65C3 used in the design SMPS indicates a nominal breakdown voltage of 650 V with a typical avalanche threshold value $V_{brds} = 730$ V for a current $I_d = 4$ A. The obtained experimental value is in accordance with technical data.

13.4.2   First step to model

Based on the obtained experimental avalanche voltage failure results, a MOSFET failure model has been built. The MOSFET failure model is built around the spice manufacturer one. The modeling of MOSFET failure consists of using switches controlled by the determined failure levels. In this way, the equivalent component state after failure (equivalent failure model in Figure 13.11) replaces the manufacturer model when the simulated voltage V_{ds} and the current I_d, for a specific time duration,

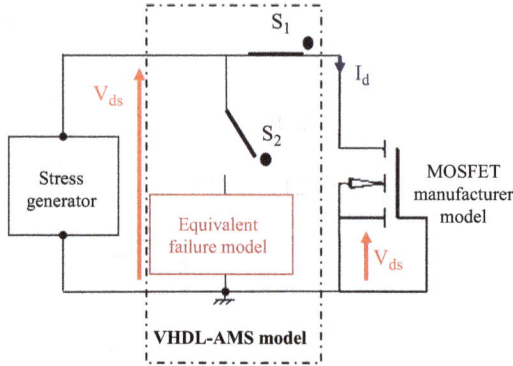

Figure 13.11   *Electrical diagram simulating the avalanche tests on MOSFET*

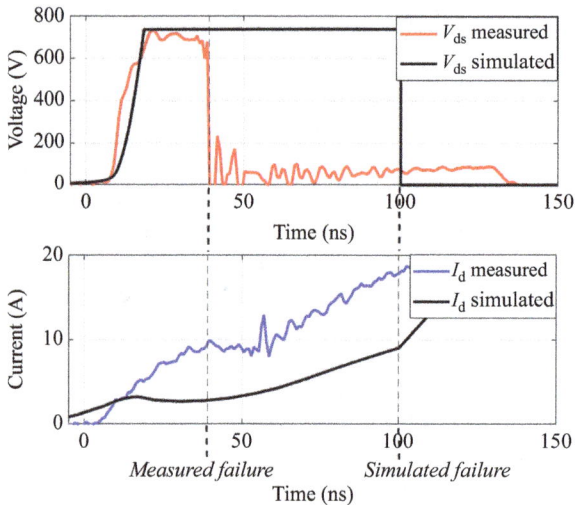

Figure 13.12   *Measurement/simulation comparison of MOSFET's voltages/ currents for avalanche tests*

are higher than the failure levels. The switches and the equivalent failure model are described in a VHDL-AMS model block that is presented in Figure 13.11.

In simulation, the pulse increases to a voltage V_{ds} of 730 V with a current I_d increasing to reach 9 A at 100 ns, failure level in current. At this moment, the voltage V_{ds} drops, and the MOSFET goes into a failed state: the manufacturer model is replaced by the equivalent model of the failed MOSFET: a short circuit of a few milliohms.

Some differences can be observed between simulated and measured current in Figure 13.12. These differences are certainly related to the impedances of the spice manufacturer model used in simulation that are not fully considered, especially in avalanche mode. This is one of the limitations of using manufacturer models in numerical simulation. However, the duration until the MOSFET failure in a power supply is about a few microseconds, the difference of a few tens of nanoseconds between the measured and the simulated failure is considered acceptable for the modeling of power supply destruction.

This section has focused on the avalanche failure modeling method of the MOSFET transistor in power supply. However, some other failure phenomena can occur in the power supply like gate overvoltage. Further work will consist of enhancing the developed component failure model.

In the next step, failure models must be developed for each destroyed component and integrated into Simplorer simulation software in order to understand and predict the SMPS behavior when a very high amplitude current pulse is injected at its input.

13.5   Conclusion

For 30 years, high-altitude nuclear explosions (NEMP/HEMP) have still been an actual concern in the field of security and safety on a national scale. In fact, NEMP couples efficiently on aerial lines of the electricity network allowing it to supply houses and factories. Once coupled to these lines, the generated interference can then be propagated to the first encountered systems and disturb or destroy them. In this context, the destruction effects of electronic power supplies due to high-level current pulse injection are studied. An advanced understanding of the power supply failure mechanisms is the key to predicting destruction. In this approach, experimental and theoretical research are implemented to predict numerically the electromagnetic interference effects on electronic power supplies. An organized approach to predict destruction effects induced by high-level current pulse injection on power supplies is developed in this chapter.

An injection generator used in this research is presented with its characteristics as well as the EUT: a flyback SMPS. For the study, a representative SMPS of a majority of common power supplies has been designed in order to fully control its topology and components. Thus, experimental studies have shown the destructive effects of these high current level pulses on power supplies designed. The main failed components responsible for the destruction of the power supply are observed: the rectifier diode, the diode bridge, and the MOSFET. Current and voltage measurements inside SMPS during pulse current injection have allowed us to confirm the destruction cause of each component and permit us to identify a sequence of component failures.

Each of these components has been individually subjected to pulses in voltage and current in order to extract the characteristics of its failures. The obtained thresholds are then used to build a failure model for each studied component. All the developed models are simulated under conditions similar to those of the experiments in order to make an objective comparison and validate each one. This chapter focuses on the MOSFET transistor, the first failed component during "struck" on the SMPS, and the methodology applied to this example.

In further work, by associating the failure models of all components, we will then be able to simulate numerically the sequence of component failures of the power supply destruction scenario and therefore predict the destruction effects induced by high-level current pulse injection on power supplies.

References

[1]   Montaño R, Bäckström M, Mansson D, *et al.* On the response and immunity of electric power infrastructures against IEMI — Current Swedish initiatives. In: *Asia-Pacific Symposium on Electromagnetic Compatibility and 19th International Zurich Symposium on Electromagnetic Compatibility*; 2008. pp. 510–513.

[2]   Radasky WA, and Hoad R. An overview of the impacts of three high power electromagnetic (HPEM) threats on Smart Grids. In: *International Symposium on Electromagnetic Compatibility – EMC EUROPE*; 2012. p. 1–6.

[3]   Foster JS, Gjelde E, Graham WR, *et al. Report of the Commission to Assess the Threat to the United States from Electromagnetic Pulse (EMP) Attack*; 2004.

[4]   Kularatna N. Powering systems based on complex ICs and the quality of utility AC source: an end to end approach to protection against transients. In: *Proceedings of Power Quality*; 2005.

[5]   Parfenov YV, Zdoukhov LN, Radasky WA, *et al.* Conducted IEMI threats for commercial buildings. *IEEE Transactions on Electromagnetic Compatibility*. 2004;46(3):404–411.

[6]   Hoad R, Lambourne A, and Wraight A. HPEM and HEMP susceptibility assessments of computer equipment. In *International Zurich Symposium on Electromagnetic Compatibility*; 2006. p. 168–171.

[7]   Girard M, Dubois T, Hoffmann P, *et al.* Effects of HPEM stress on GaAs low-noise amplifier from circuit to component scale. *Microelectronics Reliability*. 2018;88–90:914–919.

[8]   Camp M, Garbe H. Susceptibility of personal computer systems to fast transient electromagnetic pulses. *IEEE Transactions on Electromagnetic Compatibility*. 2006;48:829–833.

[9]   Nitsch D, Camp M, Sabath F, *et al.* Susceptibility of some electronic equipment to HPEM threats. *IEEE Transactions on Electromagnetic Compatibility*. 2004;46(3):380–389.

[10] Mejecaze G, Dubois T, Curos L, *et al.* Destruction analyses of power supplies due to electric pulse. *Microelectronics Reliability*. 2019;100–101.

[11] Texas Instrument. *Power Topologies Quick Reference Guide*; 2016. Available from: https://www.ti.com/lit/ug/slyu032/slyu032.pdf.

[12] ON Semiconductor. *Switch-Mode Power Supply Reference Manual*; 2014. Available from: https://www.onsemi.com/pub/collateral/smpsrm-d.pdf.

[13] Mejecaze G, Curos L, Dubois T, *et al.* Modeling of a current injection system for susceptibility study. *IEEE Transactions on Electromagnetic Compatibility*. 2020;62(6):2737–2746.

[14]   International Electrotechnical Commission. *Electromagnetic Compatibility (EMC) – Part 2-9: Environment – Description of HEMP Environment – Radiated Disturbance*. IEC 61000-2-9; 1996.

[15]   James S, Kularatna N, Steyn-Ross A, *et al.* Investigation of failure patterns of desktop computer power supplies using a lightning surge simulator and the generation of a database for a comprehensive surge propagation study. In: *Annual Conference on IEEE Industrial Electronics Society (IECON)*; 2010. p. 1275–1280.

[16]   Mejecaze G, Curos L, Dubois T, *et al.* Failure scenario of power supply due to conducted electric pulse from E1 HEMP. *IEEE Transactions on Electromagnetic Compatibility*. 2023;65(2):464–474.

[17]   Mejecaze G, Curos L, Dubois T, *et al.* Destruction scenario of power supply due to conducted pulse from HEMP. In: *2022 Global Electromagnetics (GlobalEM) Conference*. Abu Dhabi; 2022.

[18]   Kelley MD, Pushpakaran BN, Bilbao AV, Schrock JA, Bayne SB. Single-pulse avalanche mode operation of 10-kV/10A SiC MOSFET. *Microelectronics Reliability*. 2018;81:174–180.

[19]   Dchar I, Zolkos M, Buttay C, *et al.* Robustness of SiC MOSFET under avalanche conditions. In: *2017 IEEE Applied Power Electronics Conference and Exposition (APEC)*; 2017.

[20]   Curos L, Mejecaze G, Dubois T, *et al.* MOSFET failure modelling in flyback SMPS under high level conducted electrical pulses. In: *2022 Global Electromagnetics (GlobalEM) Conference*. Abu Dhabi; 2022.

Chapter 14

Electromagnetic security: threat models and exploitation scenarios exploiting susceptibility to HPEM

José Lopes Esteves[1]

This chapter discusses the existing relationship between electromagnetic compatibility (EMC) and information security (infosec) and provides an introduction to electromagnetic security (EMSEC). EMC is about assessing both the electromagnetic (EM) emissions and the susceptibility of electronic systems in order to guarantee their ability to coexist without disrupting each other. Infosec is about ensuring that sensitive information is properly protected in terms of confidentiality, integrity, and availability. EMSEC focuses on the threats to the information processed by electronic devices which are caused by EM emissions and susceptibility. A brief state-of-the-art EM security is proposed with an emphasis on the different threat models exploiting the susceptibility of electronic devices to compromise information security.

This chapter is organized as follows: in Section 14.1, the main principles of infosec are recalled, from the CIA triad to security evaluation and digital forensics and incident response. Then, Section 14.2 introduces EMSEC. Threats exploiting EM emissions are briefly recalled in Section 14.3 for the sake of completeness. In Section 14.4, threats exploiting EM susceptibility are presented in more detail, as they can be triggered by high-power electromagnetics (HPEM), showing their impact mostly on the availability or the integrity of the information processed by the target.

14.1 Information security

Infosec regroups several technical fields aiming at protecting information during its lifecycle into an information system and managing risks related to information compromise. The ISO/IEC 27000 [1] standard family provides a framework for

[1]Wireless Security Lab, Agence Nationale de la Sécurité des Systèmes d'Information (ANSSI), France

setting up and operating information security management systems. In these standards, information security is defined as follows:

Information security ensures the confidentiality, availability and integrity of information. It is achieved through the implementation of an applicable set of controls, including policies, processes, procedures, organizational structures, software and hardware to protect the identified information assets. This requires the management of risk and encompasses risks from physical, human and technology related threats with all forms of information.

The main steps of establishing an information security management system are summarized in [1]:

• Identify information assets and their associated information security requirements;
• Assess information security risks and treat information security risks;
• Select and implement relevant controls to manage unacceptable risks;
• Monitor, maintain, and improve the effectiveness of controls associated with information assets.

14.1.1   Information security requirements

The ANSSI (the national cyber security agency of France) has issued a guidance methodology for information security risk management which tries to unify several standards related to information security and risks management. Identifying information security requirements for each asset is described as determining security needs in terms of confidentiality, integrity, and availability [2]. However, those terms are rarely precisely defined in information security-related literature and can sometimes be misinterpreted or confusing. In an attempt to unify the definitions given in [1,3,4], the terms of the CIA triad can be defined as follows:

• Confidentiality is the property that information is disclosed only to authorized entities according to the security policy;
• Integrity is the property that information is complete, accurate, and has been modified only by authorized entities according to the security policy;
• Availability is the property that information is accessible and usable on demand by any authorized entity according to the security policy.

14.1.2   Security evaluation

An information system nowadays most generally involves information technology (IT) systems made of several computers or embedded systems that generate and process information and share data using communication protocols. To protect information, IT systems can enforce several protective measures which can be viewed as a set of security functions: access control, data encryption, intrusion detection, message authentication... Assessing security risks and choosing the right countermeasures requires a good understanding of the security functions implemented or deployed and more especially their limitations. In other words, for each security function, it is necessary to determine what it protects, from what kind of attacks, and in which operational conditions. It is also relevant to identify if and how the security function can be attacked. This is the main technical purpose of security evaluations of IT products.

Several methodologies for evaluating the security of IT products exist, among which the Common Criteria (CC) for Information Technology Security Evaluation [4]. The CC is a standard permitting comparability between the results of independent security evaluations.

This standard provides a security evaluation methodology [5] which defines the purpose of vulnerability analysis as determining the existence and exploitability of flaws or weaknesses in the target of evaluation in its operational environment. When a vulnerability is found, impacts on the security function and the information security requirements of the information enforced by the security function need to be assessed. Furthermore, the attack potential [5], also called the vulnerability exploitation rating [6] has to be determined. This concept reflects the so-called *attacker profile* which quantifies the strength the attacker needs to have in order to be able to exploit the vulnerability. Several criteria can be considered for this quantification, such as the attacker's expertise level, the required equipment, the time needed to identify the vulnerability and to exploit, the amount of knowledge of the target...

Generally, a maximum attack potential is defined when the scope of a security evaluation is described. It can be imposed by the standard (as in [6]), specified in the security target (document describing the evaluated perimeter in CC evaluations), or determined by the analyst according to the intended use of the product and its operational environment. The term *threat model* is also commonly used in analyses of security mechanisms. The threat model reflects the maximum attack potential considered in the analysis by defining the attacker's capabilities and the actions it is allowed to perform.

14.1.3   Digital forensics and incident response

Achieving absolute security, in the sense that the target system can resist any attacker profile and any threat (even the exploitation of unknown vulnerabilities) is not realistic. Thus, any information system can (and will probably) be compromised. Having the ability to properly respond to security incidents allows us to both limit the damage of a successful attack and also recover from the associated damage. This is the main purpose of digital forensics and incident response (DFIR) [7].

Each incident starts with the first time the targeted organization becomes aware of an event or series of events indicative of malicious activity. This step of DFIR is called the detection phase. The awareness of these events can be the result of the use of specific monitoring tools (e.g., network probes, intrusion detection systems...) or come from internal or external sources noticing suspicious activity.

After detection, the suspicious events need to be analyzed to determine if they qualify as potential incidents. This is the beginning of the analysis phase. In this phase, evidence is collected from impacted systems and examined. The ultimate goal is to determine the root cause of the incident and reconstruct the actions of the attacker.

Then comes the containment phase, aiming at restricting the ability of the attacker to further compromise the information system, and the recovery phase consists of removing the threat actor from the information system and restoring it in a safe state. The post-incident step is dedicated to the documentation of the incident and its assimilation into the infosec policies, technical measures, and organizational processes.

14.2   Electromagnetic security

EMC and infosec have very distinct goals. While in EMC the idea is to ensure a world of compliance and self-respect for all electric or electronic systems, although the immunity to the most probable electromagnetic environments is also looked for, infosec introduces an attacker in the equation which has by definition no reason to respect or comply. Furthermore, the attacker will benefit from the fact that his targets comply and follow the rules and will generally follow the weakest link to attack, which is unrelated to the most probable one.

However, phenomena corresponding to an EMC problem can be intentionally exploited by an attacker and therefore become threats for infosec. Electromagnetic security (EMSEC) can be defined as a subset of infosec considering the threats to the security of information processed by an electronic system that comes from electromagnetic interaction between the system and its electromagnetic environment. In this perspective, emission problems are considered potential threats to confidentiality, and susceptibility problems can lead to compromising the integrity or the availability of the information processed by the system [8].

This relationship is schematized in Figure 14.1.

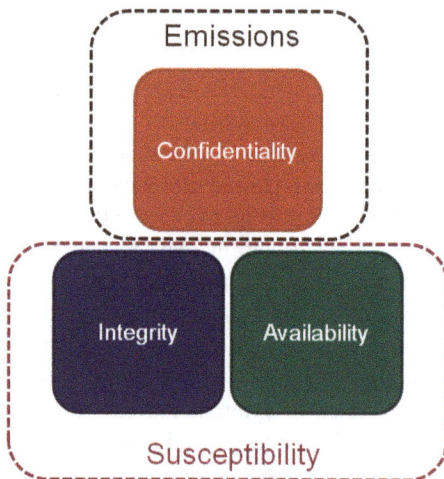

Figure 14.1   Relationship between the EMC problem and threats for infosec

14.3   Threats from electromagnetic emission

In an information system composed of electronic devices, the way information is processed (generated, displayed, stored, modified, transmitted...) involves time-varying currents, which in turn induce radiated or conducted electromagnetic emissions. If sensitive information (i.e., which has to be protected) is processed, its security might be threatened by an attacker who is able to collect the related emissions. The study of such threats can be called emissions security (and is sometimes also abbreviated EMSEC) [3, Chapter 19]. Those spurious electromagnetic emissions related to the processing of sensitive information are sometimes referred to as compromising emanations. The exploitation of such phenomena can lead to three main threats: TEMPEST, soft-tempest, and side-channel analysis.

In the *TEMPEST* threat model, a target system, during its normal operation, generates compromising emanations. The attacker is supposed to be able to collect those compromising emanations and his main goal is to reconstruct the sensitive information that is being processed.

TEMPEST attacks were demonstrated against several technologies with a focus on user interfaces, such as computer display video interfaces: cathodic ray tube displays (CRT) [9], LCD [10–12], VGA, DVI, HDMI interfaces [13], and displays from tablet computers [14]. Together with the video signal leakage, the touchscreen feature has also been considered in [15] and precise positioning of the touched spot on the vertical axis has been shown with a near field measurement. Keyboards, as a primary communication interface between the user and electronic devices, have also been scrutinized [16–20]. An RS-232 communication link has been studied in [21] and in [22], the reconstruction of a 10base-T Ethernet signal has been shown. The interaction of analog audio signals and digital I2C signals with local oscillators for radio communications enclosed in a mixed signals IC has been investigated in [23].

The *Soft-Tempest* threat model can be formulated as follows: the attacker succeeded in introducing a software implant into the target system, which generates and/or controls compromising emanations in order to modulate information on a half-duplex outbound electromagnetic covert channel. The compromising emanations can then be collected by the attacker which can reconstruct the covertly exfiltrated information.

Most Soft-Tempest studies focused on exploiting known compromising emission sources via software to create so-called exfiltration covert channels, e.g., leakage from video display units and interfaces [24–26], USB transactions [27], CPU activity [28,29], computer power supply [30]. The investigation of Soft-Tempest generated compromising emanation with radio frequency transmitters has highlighted the possibility of those parasitic signals modulating the local oscillators and being amplified by the front-end, a threat called *Second Order Soft-Tempest* [31,32]. Recent advances in Soft-Tempest have been proposed in [33] by applying 1-bit modulation techniques in order to widen the range of possible modulations for compromising emissions.

The threat model for *Side channel analysis* (SCA), from a cryptology viewpoint, is a *gray box* model, as the attacker can observe or manipulate inputs or outputs of the cryptosystem, but also get insight into the internal state of the target by analyzing its emissions [34]. The main goal is mostly restricted to cryptanalysis and recovering the secret key. Generally, the attacker is in close physical proximity with the IC performing the cryptographic computations, but recent work has also shown possible extended-range attacks exploiting parasitic propagation through network cables [35] or amplification by an RF front-end located in the same mixed-signal IC [33].

Physical cryptanalysis by measuring conducted emissions has been first proposed and applied to the DES cipher in [36]. Radiated emissions were introduced in [37], where the DES, COMP-128, and RSA primitives were attacked. Since then, attacks have been proposed against several implementations of the mostly deployed crypto-graphic primitives used for encryption, authentication, or signatures, such as ECDSA [38], AES [39], and MILENAGE [40], on different computing platforms ranging from smartcards and RFID smartcards [41] to laptops [35], and smartphones [42].

14.4   Threats from electromagnetic susceptibility

Elements composing an information system can be electric or electronic devices that intrinsically involve electric, electronic, or electromagnetic interactions to operate. They are therefore composed of several conductive parts, enclosures, power and communication cables, connectors, wires, printed circuit board (PCB) power planes and transmission lines, integrated circuit (IC) pads, bonding wires, antennas, etc. Electromagnetic fields around those conductive parts introduce parasitic currents and voltages [43]. These phenomena can be exploited by an attacker in order to introduce unwanted signals into a target device and trigger unpredicted behavior.

The introduction of parasitic signals into a system can result in impacts on the availability of the information it processes if the system is disrupted or if com-munication with other systems is made impossible. Impacts on integrity can also occur if the parasitic signals alter functional signals in a way that the information they convey is also altered and (mis)interpreted. Moreover, when processing units are disturbed, data, instructions, and execution flows can also be altered, leading to specific threats to programs enforcing security functions (Figure 14.2).

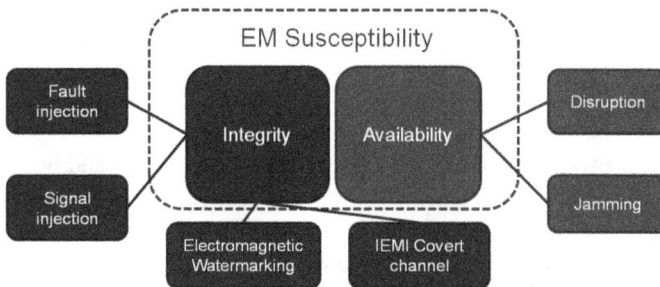

Figure 14.2   Threats for infosec related to the electromagnetic susceptibility of the target

14.4.1 Denial of service and jamming

Destructive (and permanent) effects of parasitic signals induced on systems have been reported on several targets. Thermal effects or electrical dielectric break-downs can occur and cause damage to semiconductor components and PCBs such as melted traces, bond wire destruction, and flashover effects [44]. Such effects are very likely to imply a functional failure of the whole or at least part of the system. A recent study on power supplies has confirmed those observations, showing physical damage on discrete components (e.g., resistors), IC, and PCB traces [45]. Of course, without a proper power supply, many systems may stop functioning. Physical damage to RF front-ends has also been observed, the receiver low noise amplifier (LNA) being reported as one of the weakest elements [44,46]. In that case, communication is simply made impossible. Computers have also been studied in [47] and have highlighted the possibility of peripheral damage, functional damage, or permanent damage to the tested devices. Repeatedly shutting down the target or triggering reboots can also result in a functional denial of service until the system has booted and the application environment is back up and running.

Non-destructive transient effects on conductive parts involved in communication, such as cables, PCB tracks, or RF front ends can introduce noise signals and reduce the efficiency of the communication protocol. Examples of such effects have been reported on network cables [48–50]. In the radiated case, if the noise is in the same frequency band (i.e., in-band) as the legitimate communication signal, it can be viewed as a classical RF jamming problem. In both cases, the communication is jammed during the activation of the noise source.

Finally, the possibility of exploiting functional weaknesses of power amplifiers in RF transmitters has been proposed in [51], where an attacker introduces parasitic signals during the measurement of output distortion, resulting in false pre-distortion processing and thus in temporary jamming of the output signal. In that case, the attacker signal is present only during the measurement and the jamming is effective until the next measurement phase.

14.4.2 Signal injection

The electromagnetic susceptibility of a target system can be exploited to introduce parasitic signals specifically crafted in order to get them interpreted by the system. In this threat category, the induced parasitic signal can be in-band and share the same characteristics as the expected legitimate signal or it can be out-of-band (OOB) and be transformed into or interpreted as an in-band signal by the target [52]. The possibility of inducing an RF signal into an ultrasonic front-end in order to attack an ultrasound-based distance bounding protocol has been mentioned in [53]. A formalization of such attack vectors has been proposed in [54], where in-band parasitic signals were successfully interpreted by cardiac implantable devices with severe consequences such as pacing inhibition and defibrillation. Radiated OOB Amplitude Modulation (AM) signals targeting analog microphones are shown to be first demodulated and brought in-band and then interpreted by a Bluetooth headset and a webcam, opening the possibility of exploitation scenarios such as injecting voice or noise during a phone call.

In [55–57], audio front-ends are further targeted on smartphones, both with radiated and conducted coupling modes, in order to allow an attacker to interact, remotely, silently, with integrated voice assistants. In the radiated case, the target coupling interface was a wired headphones cable, while in the conducted case, the attacker's signal was injected through the power network.

Magnetic wheel speed sensors from a car anti-lock braking system are disrupted and spoofed in [58] using a flat PCB coil to produce a magnetic field at specific frequencies.

Perturbation of sensors and actuators has been investigated in [59]. On analog sensors, previous results were confirmed showing that modulated signals can be demodulated by non-linearity in the victim circuit. On digital sensors, the study focused on the reading errors of a microcontroller general purpose input output pin while an attacker introduced continuous wave (CW) and sawtooth parasitic signals into a jumper cable. Bit flips were observed in both cases, but a deterministic control of the resulting values was better achieved with a sawtooth synchronized with the rising and falling edges of the victim circuit sampling clock. A PWM-driven servo-motor has also been analyzed while parasitic signals were injected into the cable transmitting the PWM control signal. It has been observed that a pulsed sinusoidal interference signal could introduce a DC offset leading to an increase of the PWM pulse width if injected during a high state, thus increasing the rotation angle of the servo-motor. Similarly, a sawtooth interference during a high state could introduce a parasitic falling edge, resulting in a shortening of the PWM pulse width, thus reducing the rotation angle of the servo-motor.

Capacitive touch screens from smartphones were targeted in [60] by injecting a CW signal from a copper plate hidden 5 mm behind the target. Effective signal frequencies of a few hundred kilo Hertz, which are target-dependent, introduce fake touch events that are exploited to force clicks on confirmation buttons without user interaction.

14.4.3   Fault attacks

Electromagnetic fault injection (EMFI) is a discipline dedicated to the investigation of attacks on ICs and exploiting their EM susceptibility. Conducted EMFI is often referred to as glitching and generally targets power or clock pins. Radiated EMFI is generally a near-field interaction with an EM probe placed a few millimeters over the target. These threat models imply that the attacker is assumed to have physical access to the target IC, with the possibility to adapt its environment (e.g., removing filters) and synchronize to the target activity. In most scenarios, the attacker is also able to interact with the target, provide inputs, trigger specific processing, and get access to the result, as depicted in Figure 14.3. With EMFI, attackers have been mostly impacting the integrity of data and processing, opening the way to several exploitation scenarios.

Random number generation is a cornerstone of many cryptographic protocols. True random number generators (TRNG) are components dedicated to the production of random number sequences. Using EMFI against TRNGs to bias the random number production has been proposed [61–63].

Processing corruption with EMFI has led to several exploitation scenarios benefiting from an alteration of the execution flow, such as instruction skip. As a result, privilege escalation on a rich operating system (Linux) has been shown possible in [64,65]. Attacks targeting verifications for secure boot were also demonstrated on several SoCs, allowing attackers to load unsigned firmware [66–68]. Protections present in SoCs to secure access to debugging and programming interfaces were also bypassed with EMFI [69,70]. Finally, memory leaks were also obtained with EMFI, with a corruption of a length verification during the forge of a response packet of the USB protocol, resulting in a response returning much more data than necessary [71,72].

The active counterpart to side-channel analysis consists in using EMFI to corrupt cryptographic operations to perform a cryptanalysis and recover secret keys [73]. Several cryptosystems have been considered, in both encryption and digital signature contexts, such as DES [74], AES [75], RSA [76], or elliptic curve cryptography [77].

14.4.4   Covert communication

Exploiting the electromagnetic susceptibility of a target in order to create an inbound covert communication channel was explored first in [78,79]. The threat model is the following (Figure 14.4): a software implant is running on the victim system, waiting for an external interaction in order to operate (e.g., a trigger signal,

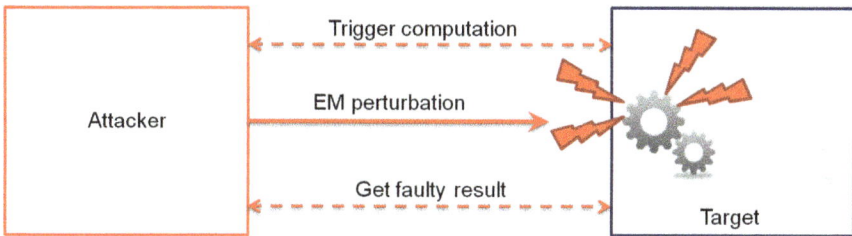

Figure 14.3   A generic threat model for EM fault injection

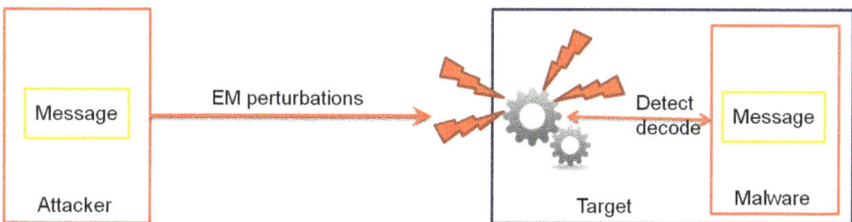

Figure 14.4   A threat model for HPEM covert channel

a software payload to execute...). The victim system is known to be susceptible to specific electromagnetic interference and the software impacts can be observed by the software implant. The attacker is able to emit a signal specifically crafted in order to modulate information on the observed software effect occurrence, so as to communicate with the software implant. Two cases are considered depending on the transmitter: it can be a software implant exploiting a transmission channel on a computer (e.g., an RF interface or an EM leakage source) or a specific RF generation equipment. In [8], an approach for determining the resulting covert channel capacity has been formalized. The result is an estimation of the channel capacity in the worst case, allowing efficient infosec risk analysis.

14.4.5   Electromagnetic watermarking

The idea of using some specific effects of HPEM on electronic systems in order to store information remotely and covertly into a non-cooperating target was explored in [80], a new approach that was called electromagnetic watermarking (EMW). Some effects of HPEM have an impact that spreads to the software (logical) layer. Such impact can be further cascading towards a modification of the logical state of the target, resulting in a storage channel. This EM storage channel can be exploited to introduce information into the target, which can be detected or extracted later when needed. Thus, EMW can be modeled as the conjunction of an EM covert channel and a storage channel (Figure 14.5).

The concept of EMW is defined and formalized. The methodology for identifying and characterizing EMW channels is further described in [8]. It involves the EM susceptibility testing of the target and the estimation of the EMW channel capacity.

An example of the application of such a technique has been implemented on a civilian UAV, which has shown to fit particularly well for such so-called electromagnetic watermarking [81]. In [82], EMW has been applied with HPEM to perform forensics tracking on UAVs, which consists in providing information about a target activity in space and time. In the case of UAVs, EMW might be considered an interesting alternative to complete neutralization against intruders, by inserting a watermark that ties the target to the intrusion event and place. This watermark can be detected and extracted afterwards, during the forensic analysis of a suspicious UAV.

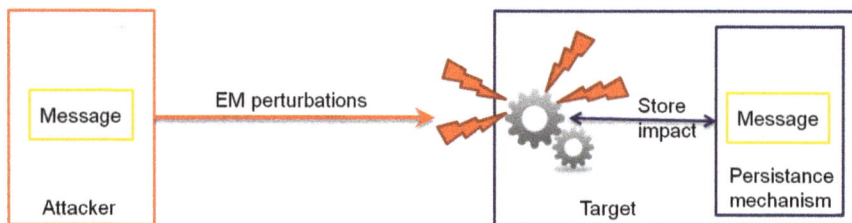

Figure 14.5   Electromagnetic watermarking: threat model

14.5   Conclusion

Information security and electromagnetic compatibility seem to be completely unrelated research and engineering topics. The former is the art of protecting sensitive information during its lifecycle against threats that are incarnated by an attacker. Security analysis consists in the identification of vulnerabilities and the determination of the impacts of their exploitation and the abilities an attacker needs to have to be able to exploit. DFIR is part of infosec which is related to the detection of attacks, live or during digital investigation, the remediation, and the recovery. EMC is the art of ensuring the peaceful and respectful coexistence of electrical and electronic systems with regard to the electromagnetic noise generated by their activity. This is achieved by controlling the emissions and the susceptibility of such systems against thresholds that reflect the most probable situations those can be confronted with.

However, when information is processed by electrical or electronic devices, the physical phenomena creating EMC problems can also become infosec threats. The study of such threats and their exploitation to compromise information security belongs to the field of electromagnetic security.

Those have been introduced in this chapter and each corresponding threat model has been explained in order to give an understanding of how an attacker can exploit the susceptibility of a device in order to compromise the security of the information processed. Impacts on information availability can be achieved mostly by jamming or physical destruction. Integrity can be impacted and lead to signal injection and fault injection.

For threats involving the exploitation of the susceptibility of a target, a security analysis requires a susceptibility analysis considering electromagnetic environments reflecting an attacker. EMC standards defined such electromagnetic environments as IEMI [83]. In infosec, most effort to understand and take those threats into account came from studies on fault injection and sensor security. Interesting leads for the definition of IEMI or HPEM attacker profiles are explored in [84], especially regarding the source's characteristics. This characterization could also lead to detection techniques which is a fundamental necessity for accounting EMSEC threats in DFIR [47,85,86].

References

[1]   ISO/IEC. Information Technology — Security Techniques — Information Security Management Systems. International Organization for Standardization; 2018. ISO/IEC 27000:2018.

[2]   ANSSI. *EBIOS: Expression Des Besoins et Identification Des Objectifs de Sécurité – Méthode de Gestion Des Risques*. Paris, France: Agence Nationale de la Sécurité des Systèmes d'Information; 2010.

[3]   Anderson RJ. *Security Engineering: A Guide to Building Dependable Distributed Systems*. 2nd ed. New York: Wiley; 2020.

[4] Criteria C. Common Criteria for Information Technology Security Evaluation - Part 1: Introduction and General Model; 2017. Version 3.1 Revision 5.

[5] Criteria C. Common Methodology for Information Technology Security Evaluation - Evaluation Methodology; 2017. Version 3.1 Revision 5.

[6] ANSSI. Criteria for Evaluation in View of a First Level Security Certification. ANSSI; 2020. ANSSI-CSPN-CER-P-02_v4.0.

[7] Johansen G. *Digital Forensics and Incident Response: Incident Response Techniques and Procedures to Respond to Modern Cyber Threats*, 2nd ed. Birmingham: Packt Publishing; 2020.

[8] Lopes Esteves J. *Electromagnetic Interferences and Information Security: Characterization, Exploitation and Forensic Analysis*. HESAM Université. Paris, France; 2023.

[9] van Eck W. Electromagnetic Radiation from Video Display Units: An Eavesdropping Risk? *Computers & Security*. 1985;4(4):269–286.

[10] Kuhn MG. *Compromising Emanations: Eavesdropping Risks of Computer Displays*. University of Cambridge, Computer Laboratory; 2003. UCAM-CL-TR-577.

[11] Kuhn MG. Compromising Emanations of LCD TV Sets. *IEEE Transactions on Electromagnetic Compatibility*. 2013;55(3):564–570.

[12] Kuhn MG. Electromagnetic Eavesdropping Risks of Flat-panel Displays. *In: Proceedings of the 4th International Conference on Privacy Enhancing Technologies*. PET'04. Berlin: Springer; 2005. pp. 88–107.

[13] Ricordel PM, and Duponchelle E. Risques Associés Aux Signaux Parasites Compromettants: Le Cas Des Câbles DVI et HDMI. In: *Symposium Sur La Sécurité Des Technologies de l'Information et Des Communications (SSTIC)*. Rennes, France; 2018.

[14] Hayashi Y, Homma N, Miura M, *et al.* A Threat for Tablet PCs in Public Space: Remote Visualization of Screen Images Using EM Emanation. In: *Proceedings of the 2014 ACM SIGSAC Conference on Computer and Communications Security*. CCS'14. New York, NY: ACM; 2014. pp. 954–965.

[15] Hoad R. Identifying Some Radiated EMSEC Vulnerabilities of Tablet Personal Computers. In: *European Electromagnetics International Symposium EUROEM 2016*. London, UK; 2016.

[16] Vuagnoux M, and Pasini S. Compromising Electromagnetic Emanations of Wired and Wireless Keyboards. In: *Proceedings of the 18th Conference on USENIX Security Symposium*. SSYM'09. Berkeley, CA: USENIX Association; 2009. pp. 1–16.

[17] Du YL, Lu YH, and Zhang JL. Novel Method to Detect and Recover the Keystrokes of Ps/2 Keyboard. *Progress In Electromagnetics Research*. 2013; 41:151–161.

[18] Vuagnoux M, and Pasini S. An Improved Technique to Discover Compromising Electromagnetic Emanations. In: 2010 *IEEE International Symposium on Electromagnetic Compatibility*; 2010. p. 121–126.

[19] Nowosielski L, and Wnuk M. Compromising Emanations from USB 2 Interface. *In: PIERS Proceedings*; 2014.

[20]   Przesmycki R, and Nowosielski L. USB 3.0 Interface in the Process of Electromagnetic Infiltration. In: 2016 *Progress in Electromagnetic Research Symposium (PIERS)*; 2016. p. 1019–1023.

[21]   Smulders P. The Threat of Information Theft by Reception of Electromagnetic Radiation from RS-232 Cables. *Computers & Security*. 1990;9(1): 53–58.

[22]   Schulz M, Klapper P, Hollick M, *et al.* Trust The Wire, They Always Told Me!: On Practical Non-Destructive Wire-Tap Attacks Against Ethernet. In: *WiSec'16. Proceedings of the 9th ACM Conference on Security & Privacy in Wireless and Mobile Networks*. New York, NY: Association for Computing Machinery; 2016. pp. 43–48.

[23]   Choi J, Yang HY, and Cho DH. TEMPEST Comeback: A Realistic Audio Eavesdropping Threat on Mixed-signal SoCs. In: *Proceedings of the 2020 ACM SIGSAC Conference on Computer and Communications Security. CCS'20*. New York, NY: Association for Computing Machinery; 2020. pp. 1085–1101.

[24]   Kuhn MG, and Anderson RJ. Soft Tempest: Hidden Data Transmission Using Electromagnetic Emanations. In: *International Workshop on Information Hiding*. Berlin: Springer; 1998. pp. 124–142.

[25]   Thiele E. *Tempest for Elisa*; 2001. www.erikyyy.de/tempest.

[26]   Guri M, Kedma G, Kachlon A, *et al.* AirHopper: Bridging the Air-Gap between Isolated Networks and Mobile Phones Using Radio Frequencies. In: *2014 9th International Conference on Malicious and Unwanted Software: The Americas (MALWARE)*; 2014. p. 58–67.

[27]   Guri M, Monitz M, and Elovici Y. USBee: Air-gap Covert-Channel via Electromagnetic Emission from USB. In: 2016 *14th Annual Conference on Privacy, Security and Trust (PST)*. Piscataway, NJ: IEEE; 2016. pp. 264–268.

[28]   Guri M. MAGNETO: Covert Channel between Air-Gapped Systems and Nearby Smartphones via CPU-generated Magnetic Fields. *Future Generation Computer Systems*. 2021;115:115–125.

[29]   Guri M, Zadov B, and Elovici Y. ODINI: Escaping Sensitive Data From Faraday-Caged, Air-Gapped Computers via Magnetic Fields. *IEEE Transactions on Information Forensics and Security*. 2020;15:1190–1203.

[30]   Guri M, Zadov B, Bykhovsky D, *et al.* PowerHammer: Exfiltrating Data From Air-Gapped Computers Through Power Lines. *IEEE Transactions on Information Forensics and Security*. 2020;15:1879–1890.

[31]   Lopes Esteves J, Cottais E, and Kasmi C. Second Order Soft Tempest: From Internal Cascaded Electromagnetic Interactions to Long Haul Covert Channels. In: *Radio Science Conference (URSI AP-RASC), 2019 URSI Asia Pacific*. New Delhi: IEEE; 2019.

[32]   Lopes Esteves J, Cottais E, and Kasmi C. Second Order Soft-Tempest in RF Front-Ends: Design and Detection of Polyglot Modulations. In: *2018 International Symposium On Electromagnetic Compatibility-EMC EUROPE*. Amsterdam: IEEE; 2018.

[33]   Camurati G. Security Threats Emerging from the Interaction between Digital Activity and Radio Transceivers; 2020.

[34]   Lomne V. Power and Electro-Magnetic Side-Channel Attacks: Threats and Countermeasures [These de Doctorat]. Montpellier 2; 2010.

[35]   Genkin D, Pipman I, and Tromer E. Get Your Hands off My Laptop: Physical Side-Channel Key-Extraction Attacks on PCs. *Journal of Cryptographic Engineering.* 2015;5(2):95–112.

[36]   Kocher P, Jaffe J, and Jun B. Differential Power Analysis. In: Wiener M (eds.), *Advances in Cryptology — CRYPTO'99. Lecture Notes in Computer Science.* Berlin: Springer; 1999. pp. 388–397.

[37]   Gandolfi K, Mourtel C, and Olivier F. Electromagnetic Analysis: Concrete Results. In: Koç ÇK, Naccache D, Paar C (ed), Cryptographic Hardware and Embedded Systems — CHES 2001. *Lecture Notes in Computer Science.* Berlin: Springer; 2001. p. 251–261.

[38]   Roche T, Lomné V, Mutschler C, *et al.* A Side Journey to Titan. In: *30th USENIX Security Symposium (USENIX Security 21).* USENIX Association; 2021.

[39]   Ronen E, O'Flynn C, Shamir A, *et al. IoT Goes Nuclear: Creating a ZigBee Chain Reaction*; 2016. 1047.

[40]   Devine C, San Pedro M, and Thillard A. A Practical Guide to Differential Power Analysis of USIM Cards. *In: Symposium Sur La Sécurité Des Technologies de l'Information et Des Communications (SSTIC).* Rennes, France; 2018.

[41]   Oswald D, and Paar C. Breaking Mifare DESFire MF3ICD40: Power Analysis and Templates in the Real World. In: *International Workshop on Cryptographic Hardware and Embedded Systems.* Berlin: Springer; 2011. pp. 207–222.

[42]   Lisovets O, Knichel D, Moos T, *et al.* Let's Take It Offline: Boosting Brute-Force Attacks on iPhone's User Authentication through SCA. *IACR Trans Cryptogr Hardw Embed Syst. 2021*;2021(3):496–519.

[43]   Schmitt R. *Electromagnetics Explained: A Handbook for Wireless/RF, EMC, and High-Speed Electronics.* EDN Series for Design Engineers. Amsterdam: Elsevier Science; 2002.

[44]   Giri D, Hoad R, and Sabath F. *High-Power Electromagnetic Effects on Electronic Systems.* London: Artech House; 2020.

[45]   Mejecaze G. *Analyse des destructions d'alimentations électroniques soumises à un courant impulsionnel fort niveau.* Université de Bordeaux; 2019.

[46]   Girard M. *Recherche de Vulnérabilités Des Étages de Réception Aux Agressions Électromagnétiques de Forte Puissance*: Cas d'un LNA AsGa. Bordeaux; 2018.

[47]   Hoad R. *The Utility of Electromagnetic Attack Detection to Information Security.* University of Glamorgan; 2007.

[48]   Backstrom MG, and Lovstrand KG. Susceptibility of Electronic Systems to High-Power Microwaves: Summary of Test Experience. *IEEE Transactions on Electromagnetic Compatibility.* 2004;46(3):396–403.

[49] Kreitlow M, Sabath F, and Garbe H. A Test Method for Analysing Disturbed Ethernet Data Streams. *Advances in Radio Science* 2015;2015(13):149–153.

[50] Kreitlow M, Sabath F, and Garbe H. Analysis of IEMI Effects on a Computer Network in a Realistic Environment. In: 2015 *IEEE International Symposium on Electromagnetic Compatibility (EMC)*; 2015. pp. 1063–1067.

[51] Cottais E, Lopes Esteves J, Houchouas V, *et al.* Effects of Intentional Electromagnetic Interference on an Adaptive Predistortion Algorithm. *In: 2017 International Symposium On Electromagnetic Compatibility-EMC EUROPE*. Angers, France: IEEE; 2017. pp. 1–6.

[52] Giechaskiel I, and Rasmussen KB. SoK: Taxonomy and Challenges of Out-of-Band Signal Injection Attacks and Defenses. arXiv:190106935 [cs]. 2019.

[53] Rasmussen KB, Castelluccia C, Heydt-Benjamin TS, *et al.* Proximity-Based Access Control for Implantable Medical Devices. In: *Proceedings of the 16th ACM Conference on Computer and Communications Security. CCS'09*. New York, NY: Association for Computing Machinery; 2009. pp. 410–419.

[54] Kune DF, Backes J, Clark SS, *et al.* Ghost Talk: Mitigating EMI Signal Injection Attacks against Analog Sensors. In: 2013 *IEEE Symposium on Security and Privacy*; 2013. p. 145–159.

[55] Kasmi C, and Lopes Esteves J. IEMI Threats for Information Security: Remote Command Injection on Modern Smartphones. *IEEE Transactions on Electromagnetic Compatibility*. 2015;57(6):1752–1755.

[56] Kasmi C, and Lopes Esteves J. IEMI and Smartphone Security: A Smart Use of Front Door Coupling for Remote Command Execution. In: *Asia Electromagnetics Symposium (ASIAEM 2015)*. Jeju-si, Jeju Province, South Korea; 2015.

[57] Lopes Esteves J, and Kasmi C. Remote and Silent Voice Command Injection on a Smartphone through Conducted IEMI: Threats of Smart IEMI for Information Security. *System Design and Assessment Note SDAN*. 2018;48.

[58] Shoukry Y, Martin P, Tabuada P, *et al.* Non-Invasive Spoofing Attacks for Anti-lock Braking Systems. In: Bertoni G, Coron JS (eds.) *Cryptographic Hardware and Embedded Systems - CHES 2013. Lecture Notes in Computer Science*. Berlin: Springer; 2013. pp. 55–72.

[59] Selvaraj J. *Intentional Electromagnetic Interference Attack on Sensors and Actuators*. Iowa State University; 2018.

[60] Maruyama S, Wakabayashi S, and Mori T. Tap 'n Ghost: A Compilation of Novel Attack Techniques against Smartphone Touchscreens. In: *2019 IEEE Symposium on Security and Privacy (SP)*; 2019.

[61] Haddad P. Caractérisation et Modélisation de Générateurs de Nombres Aléatoires Dans Les *Circuits Intégrés Logiques*. Université Jean Monnet. Saint Etienne; 2015.

[62] Haddad P, Kasmi C, Lopes Esteves J, *et al.* Electromagnetic Harmonic Attack on Transient Effect Ring Oscillator Based True Random Generator. In: *Hardwear.Io*. The Hague, Netherlands; 2016.

[63] Madau M, Agoyan M, Balasch J, *et al.* The Impact of Pulsed Electromagnetic Fault Injection on True Random Number Generators. In: *2018

Workshop on Fault Diagnosis and Tolerance in Cryptography (FDTC); 2018. p. 43–48.

[64] Gaine C, Aboulkassimi D, Pontié S, *et al.* Electromagnetic Fault Injection as a New Forensic Approach for SoCs. In: 2020 *IEEE International Workshop on Information Forensics and Security (WIFS)*; 2020. p. 1–6.

[65] Timmers N, and Mune C. Escalating Privileges in Linux Using Voltage Fault Injection. In: 2017 *Workshop on Fault Diagnosis and Tolerance in Cryptography (FDTC)*; 2017. p. 1–8.

[66] Results L. Fatal Fury On ESP32 Time To Release HW Exploits. *In: Black Hat Europe 2019*. London, UK; 2019.

[67] Plutoo, derrek, naehrwert. Console Security – Switch. In: *34th Computer Chaos Club Congress*. Leipzig, Germany; 2017.

[68] Timmers N, Spruyt A, and Witteman M. Controlling PC on ARM Using Fault Injection. In: *2016 Workshop on Fault Diagnosis and Tolerance in Cryptography, FDTC 2016*, Santa Barbara, CA, August 16, 2016. IEEE Computer Society; 2016. p. 25–35.

[69] Bozzato C, Focardi R, and Palmarini F. Shaping the Glitch: Optimizing Voltage Fault Injection Attacks. *IACR Transactions on Cryptographic Hardware and Embedded Systems*. 2019; p. 199–224.

[70] Results L. Debug Resurrection On nRF52 Series. In: *Black Hat Europe 2020*; 2020.

[71] O'Flynn C. MIN()Imum Failure: EMFI Attacks against USB Stacks. *In: 13th USENIX Workshop on Offensive Technologies (WOOT 19)*. Santa Clara, CA: USENIX Association; 2019.

[72] Scott M. A USB Glitching Attack. In: *PoC—GTFO, 2(0×13)*. No. 2 in PoC—GTFO, 2016. p. 30–37.

[73] Rivain M. Differential Fault Analysis on DES Middle Rounds. In: *Proceedings of the 11th International Workshop on Cryptographic Hardware and Embedded Systems. CHES'09*. Berlin: Springer; 2009. p. 457–469.

[74] Biham E, and Shamir A. Differential Fault Analysis of Secret Key Cryptosystems. In: Kaliski BS (eds.) *Advances in Cryptology—CRYPTO'97*. Berlin: Springer; 1997. p. 513–525.

[75] Dehbaoui A, Dutertre JM, Robisson B, *et al.* Investigation of Near-Field Pulsed EMI at IC Level. In: 2013 *Asia-Pacific Symposium on Electromagnetic Compatibility (APEMC)*; 2013. p. 1–4.

[76] Boneh D, DeMillo RA, and Lipton RJ. On the Importance of Checking Cryptographic Protocols for Faults (Extended Abstract). In: Fumy W (ed), *Advances in Cryptology – EUROCRYPT'97, International Conference on the Theory and Application of Cryptographic Techniques, Konstanz, Germany, May 11-15, 1997, Proceeding*. vol. 1233 of *Lecture Notes in Computer Science*. Berlin: Springer; 1997. pp. 37–51.

[77] Biehl I, Meyer B, and Müller V. Differential Fault Attacks on Elliptic Curve Cryptosystems. In: Bellare M (ed), *Advances in Cryptology — CRYPTO 2000*. Berlin: Springer; 2000. pp. 131–146.

[78] Kasmi C, Lopes Esteves J, and Valembois P. Susceptibility Testing for Detecting IEMI-based Covert Channels. *In: European Electromagnetics International Symposium EUROEM 2016*. London, UK; 2016.

[79] Kasmi C, Lopes Esteves J, and Valembois P. Air-Gap Limitations and Bypass Techniques: "Command and Control" Using Smart Electromagnetic Interferences. *The Journal on Cybercrime & Digital Investigations*. 2016;1(1): 13–19.

[80] Lopes Esteves J. Electromagnetic Watermarking: Exploiting IEMI Effects for Forensic Tracking of UAVs. In: *2019 International Symposium on Electromagnetic Compatibility-EMC EUROPE*. Barcelona: IEEE; 2019.

[81] Lopes Esteves J. Active Forensics Tracking Exploiting Logical Effects of HPEM. In: *General Assembly and Scientific Symposium (URSI GASS), 2020 XXXIIIrd URSI*. Rome, Italy; 2020.

[82] Lopes Esteves J, Cottais E. Covert Information Embedding in Remote Targets with HPEM. In: *Asia Electromagnetics Symposium (ASIAEM 2019)*. Xian, China: Summa Foundation; 2019.

[83] ISO/IEC. *Electromagnetic Compatibility (EMC) – Part 4-36: Testing and Measurement Techniques – IEMI Immunity Test Methods for Equipment and Systems*. Geneva: International Organization for Standardization; 2020. ISO/IEC 61000-4-36:2020.

[84] Mora Parra N. *Contribution to the Study of the Vulnerability of Critical Systems to Intentional Electromagnetic Interference (IEMI)*. EPFL. Lausanne; 2016.

[85] Hoad R, and Sutherland I. The Forensic Utility of Detecting Disruptive Electromagnetic Interference. In: *ECIW2008 – 7th European Conference on Information Warfare and Security: ECIW2008*. Academic Conferences Limited; 2008. p. 77.

[86] Lopes Esteves J. Agressions Electromagnétiques et Forensics. In: *Conférence Sur La Réponse Aux Incidents et l'investigation Numérique (CoRI&IN) 2019*. Lille, France: Cecyf; 2019.

Chapter 15

Development of a 1D and 2D slotted waveguide antenna arrays in S-band for multi-megawatt applications

Zubair Akhter[1], Evgeny Gurnevich[1], Abdulla AlAli[1], Sundhar Venugopal[1], Mariam Almenhali[1], Luciano Prado[1], Ernesto Neira[1], Mae AlMansoori[1], Fernando Albarracin[1], Felix Vega[1] and Chaouki Kasmi[1]

This chapter introduces an innovative antenna design known as the High-Power Slotted Waveguide Antenna Array (HPSWA), specifically engineered for high-power electromagnetic (HPM) generators. The design and development of the 1D and 2D-HPSWA have been driven by the increasing demand for efficient and robust antenna systems capable of handling very high power levels while maintaining a low form factor and system integrity. The HPSWA has been constructed to withstand internal pressures of up to 2 bars, utilizing sulfur hexafluoride (SF6) as a pressurizing medium. The use of SF6 not only enhances the dielectric strength properties of the inner waveguide sections but also contributes to the overall thermal stability of the antenna. Pneumatic and electrical testing has been conducted on the HPSWA to evaluate its performance characteristics comprehensively. The most relevant antenna parameters, including impedance response, gain, radiation pattern, and peak radiated power when connected to an HPM generator are presented. The developed HPSWA represents a substantial advancement in antenna technology, offering an effective means of transmitting peak power in excess of 4 MW at frequencies in S-band. The combination of high gain and the ability to operate under pressurized conditions not only enhances its applicability across various domains but also addresses critical challenges faced in the field of directed energy applications and intentional electromagnetic interference, IEMI, studies.

15.1 Introduction

A slotted waveguide antenna (SWA) consists of a hollow metallic structure (the waveguide) with one or more slots cut into its surface. These slots allow electromagnetic

[1]Directed Energy Research Center, Technology Innovation Institute, Abu Dhabi, United Arab Emirates

waves to be radiated into free space while guiding the energy along the waveguide. The design of the slots determines the antenna's radiation pattern, impedance, and bandwidth [1]. The use of waveguides minimizes losses associated with traditional coaxial cables, making them suitable for high-power applications where efficiency is crucial.

SWAs are often the preferred choice for high-power applications due to several key advantages:

1. **High Power Handling Capability:** Waveguides are inherently capable of handling high power levels without suffering from dielectric breakdown or overheating, making them suitable for applications that require the transmission of significant amounts of energy.
2. **Low Loss:** Waveguides exhibit lower transmission losses compared to traditional coaxial cables or other types of antennas, particularly at microwave frequencies. This is crucial in high-power applications, as it ensures that more of the input power is effectively radiated rather than lost as heat.
3. **Efficient Radiation Patterns:** SWAs can be designed to produce highly directional radiation patterns, which is essential for applications such as radar and communication systems that require focused energy beams.
4. **Mechanical Robustness:** The solid structure of waveguides provides better mechanical stability compared to other antenna types. This robustness is beneficial in high-power environments where physical stress and thermal expansion can impact performance.
5. **Compact Design:** SWAs can be designed to be relatively compact while still providing high gain compared to reflector-based antennas. This is advantageous for applications where space is limited or where a low-profile design is desired.
6. **Reduced/Controlled Side Lobes levels:** The design of SWAs can minimize side lobes in the radiation pattern and even optimize for specific side lobe levels (SLL), which helps reduce interference from external radiators and spillover of EM radiation in undesired directions which improves overall system performance.
7. **Thermal Management:** The metallic structure of waveguides facilitates effective heat dissipation, which is critical when operating at high power levels, helping to maintain performance and reliability.

These advantages make SWAs particularly suitable for high-power applications in fields such as radar, satellite communications, IEMI studies, and directed energy systems, where efficiency, reliability, and performance are paramount.

15.2   Proposed 1D HPSWA

High-power microwave antennas are crucial components of modern communication, IEMI research, and radar systems. One of the key features of high-power microwave antennas is their ability to handle large amounts of power without degrading the signal. Reflector [2], helical [3], waveguide horn [4,5,6], and SWA are among the preferred choices for high-power applications [7,8]. Dielectric breakdown is a critical consideration in high-power applications. When the electric field surpasses the dielectric strength of a material, it can result in the failure of the dielectric, causing it to momentarily act as a conductor. This phenomenon can lead to insulation loss and increase the risk of electrical arcing, potentially damaging the equipment. Consequently, it is essential to comprehend the dielectric strength of materials and to design systems that mitigate the risk of breakdown in high-power environments [9].

15.2.1   Conventional SWA design

The proposed high-power SWA is designed for operation within the S-band frequency range. It employs a standard WR-340 waveguide, characterized by dimensions of $a = 86.36$ mm and $b = 43.18$ mm [10], as the feeding structure of the antenna. To achieve a high-power rating, the antenna is covered with a low-loss dielectric layer, to contain a pressurized gas to ultimately implement an inner medium whose dielectric strength level is higher than that from air. This modification significantly improves the hold-off voltage rating in comparison to air. The antenna is pressurized with sulfur hexafluoride (SF6) at 2 bar and maintained at a temperature of 22 °C.

A three-dimensional model of the SWA is illustrated in Figure 15.1. The antenna is constructed with a WR-340 waveguide featuring a wall thickness of 2 mm. Each slot is designed with a length $Ls = 35.3$ mm and a width $Ws = 8$ mm, with adjacent separation among the slots $\sim \lambda_g/2$, at 3 GHz, and a uniform offset of approximately 13.5 mm from the centerline of the waveguide. Detailed design guidelines for SWA for achieving specific SLL can also be referred to elsewhere [11] in the literature. Furthermore, each slot is carefully modeled to eliminate sharp corners and edges, thereby reducing field enhancement to reduce undesirable breakdowns caused by rough or irregular surfaces (as shown in the zoomed view of the slot in Figure 15.1).

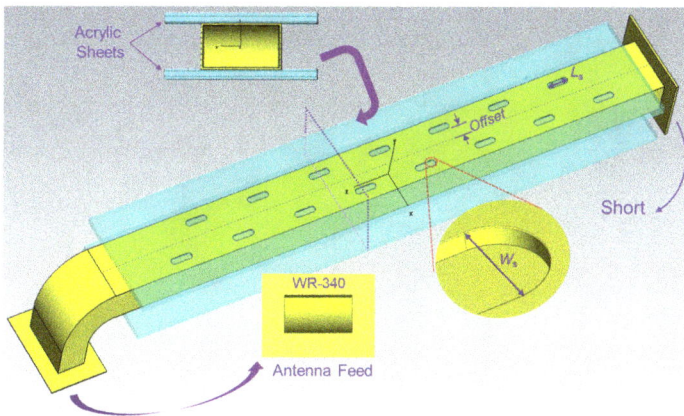

Figure 15.1   3D electromagnetic model for proposed pressurized high power slotted waveguide antenna (HPSWA)

The High-Power Slotted Waveguide Antenna Array (HPSWA) is simulated in a commercial 3D electromagnetic solver, and its reflection coefficient is shown in Figure 15.2. The antenna shows good impedance matching starting at 2.91 GHz till 3.1 GHz (bandwidth, $BW = 190$ MHz), and its resonance frequency is at 2.974 GHz with a reflection coefficient of -48.7 dB. The simulated radiation pattern is shown in Figure 15.3 and its peak realized gain at 3 GHz is about 16.6 dBi. The observed

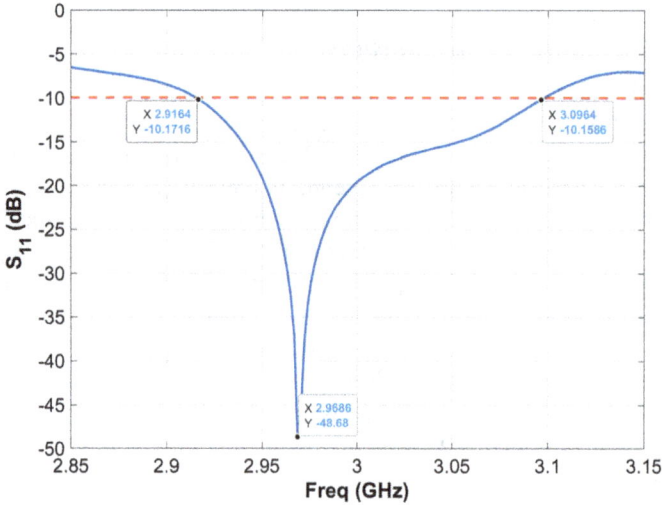

Figure 15.2   Simulated reflection coefficient

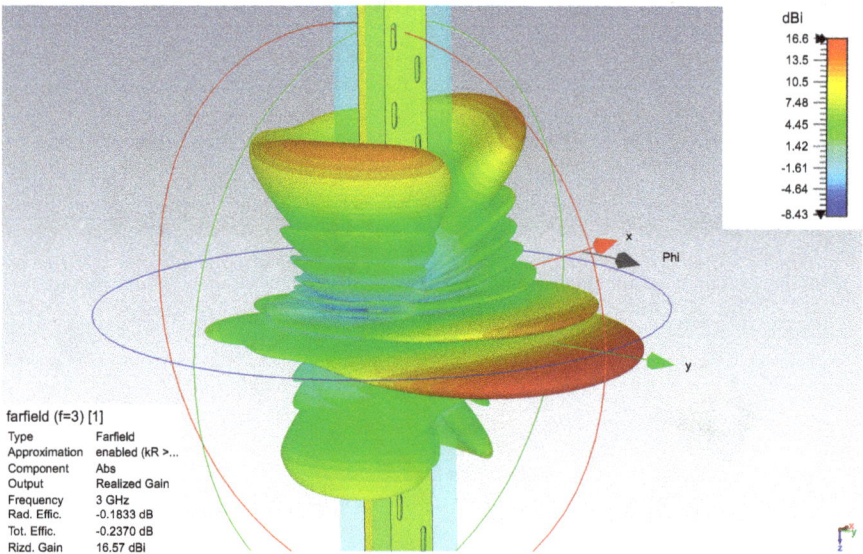

Figure 15.3   Simulated 3D radiation pattern at 3 GHz

−3 dB beam widths in the simulation are close to 6° and 54° in elevation and azimuth directions, respectively.

15.2.2 Antenna fabrication, assembly, and testing

The HPSWA's first prototype was manufactured in-house with the help of a metal laser cutter machine. A commercially available standard WR-340 straight wave-guide section was used as a base in the design. Cutting the slots on one wall of the waveguide without affecting the opposite wall, as well as obtaining a precise slots positioning along the waveguide comprised the main challenges in the process. To protect the opposite wall of the waveguide, the power of the laser was optimized and additional protective bars of different materials such as wood, metal, and wet sand were filled into the waveguide cross-section. The acrylic covering layer is pneumatically sealed with the help of a rubber gasket and silicon-based sealant as shown in Figure 15.4. The antenna is pressurized with the help of a waveguide pressure window with an integrated gas inlet point. The antenna was tested for its return-loss performance before conducting the radiation pattern measurement. The antenna return loss was measured under three conditions i.e., an antenna radiating

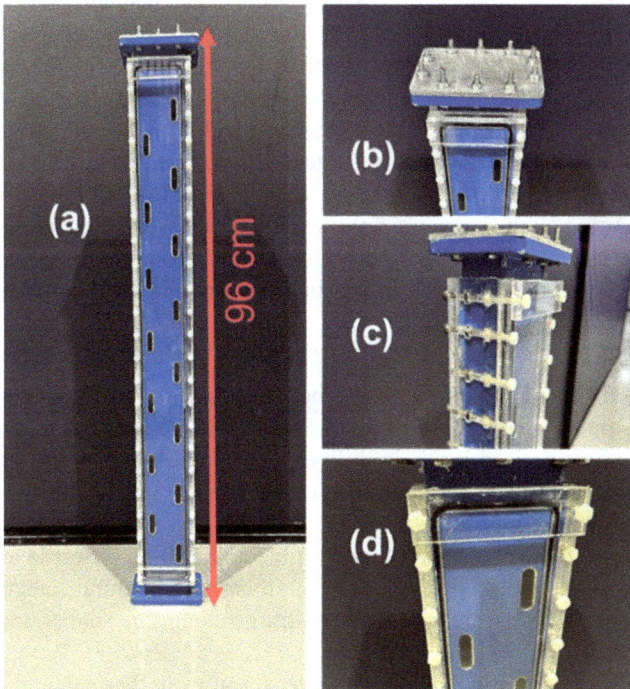

Figure 15.4   *Manufactured HPSWA. (a) Hermetically sealed walls. (b) Shorting plate with gasket. (c) Mounting of acrylic sheets and sealing mechanism. (d) Gasket placement on slotted aperture.*

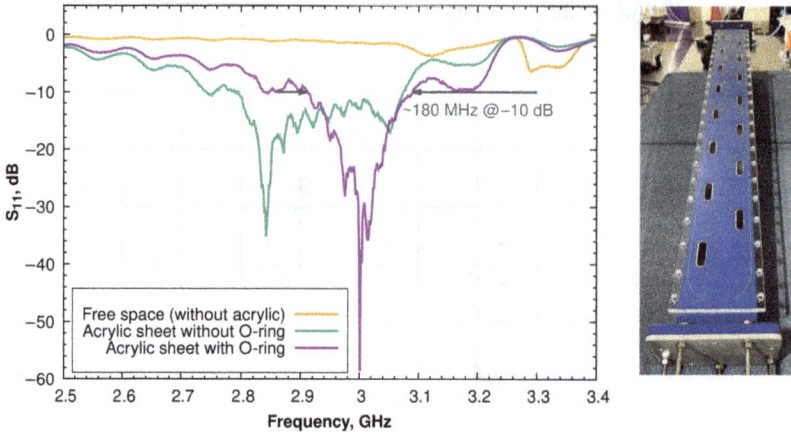

Figure 15.5   Measured reflection coefficient in free space and with acrylic sheet sealing

in free space, an acrylic sheet without an O-ring (gasket), and an acrylic sheet with an O-ring. As can be seen in Figure 15.5, the impedance matching level at 3 GHz is poor for the uncovered aperture case. If the slotted aperture is loaded with the acrylic sheet without an O-ring (i.e., there is absolutely no gap between the slots and the acrylic sheet), the antenna resonates well at 2.84 GHz. As this antenna was optimized for O-ring presence, it resonates very well and the reflection coefficient is well below −30 dB at 3 GHz. This is due to the optimal loading at the waveguide slots due to the presence of dielectric material in the close vicinity. Note that the presence of O-ring slightly increases the separation between the radiating slots and the acrylic sheet (~0.15–0.25 mm) which shifts the resonance frequency towards a higher value due to the reduction in the dielectric loading of the radiating slots.

15.2.3   Structural integrity testing under positive pressure

As highlighted before, the SWAs have become essential components in various high-power applications, particularly in radar and telecommunications. These antennas are characterized by their ability to provide high-gain and narrow beam widths, making them ideal for applications that require precise beam steering and efficiency. However, the operational integrity of these antennas under extreme conditions is paramount, especially when subjected to high power levels. Pressure testing is a critical process that ensures the antenna's structural integrity, leakage prevention, and safety assurance for the instruments and personnel under high RF power regimes.

High-power applications often involve significant thermal and mechanical stresses. Pressure testing assesses the structural integrity of the antenna under simulated operational conditions, ensuring that it can withstand the forces exerted during operation. Under real scenarios, SWAs may be exposed to harsh environmental conditions, including moisture and contaminants. Pressure testing helps

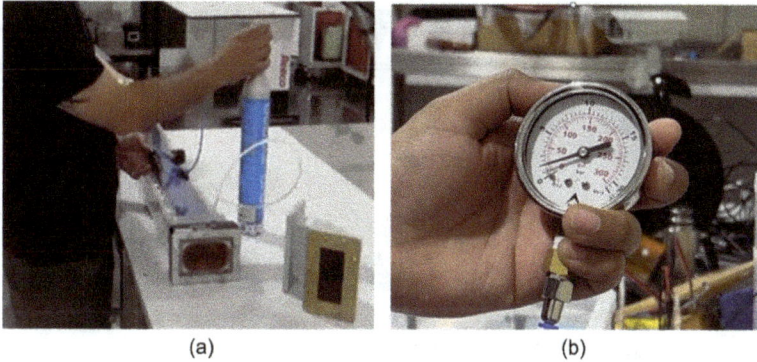

(a)   (b)

*Figure 15.6   Pneumatic testing setup (a) HPSWA under positive pressure and (b)
the reading on the pressure gauge*

identify any potential leaks in the waveguide structure, which could lead to performance degradation or failure due to dielectric breakdown.

The presented HPSWA antenna was prepared for pneumatic testing by ensuring that all connections were secure and the necessary instruments such as pressure
gauge, standard valves, piping, and pressure source were in place. An illustration of
the setup is shown in Figure 15.6. The pressure level was progressively increased
from 1 bar to 2.5 bar, in steps of 0.5 bar every 10 min to observe the response of the
antenna structure. Throughout the testing process, key parameters such as structural
deformation, and leaks were monitored. A satisfactory pneumatic and structural
performance was obtained for an acrylic plate, 6 mm thick, fixed with nylon bolts
separated 35 mm from each other (see Figure 15.4(a) for reference).

15.2.4   Radiation pattern measurement

The gain and radiation pattern of the proposed HPSWA prototype was conducted in
the semianechoic chamber at DERC-TII facilities. The scheme of the measurement
setup is shown in Figure 15.7(a). Extra absorbing material was arranged to cover the
ground surrounding the antenna setup, as can be seen in Figure 15.7(b). The process
begins with the proper placement of antennas (AUT and reference antenna) with the
help of dielectric masts and a laser alignment system. A calibrated reference antenna is
positioned 12 m away from the antenna under test (AUT), ensuring that measurements
occur in the far-field region at 3 GHz, where the radiation pattern can be accurately
measured. The AUT was mounted on a motorized turntable, which allows a synchronized continuous measurement of the S-parameters from the vector network analyzer, VNA. After completing the rotation, the collected data is processed to generate
radiation patterns. Figure 15.8 shows the radiation patterns of the E-plane
(azimuth) and the H-plane (elevation). Figure 15.9 presents a Cartesian plot of the
E- and H-planes, including the highest cross-polarization level obtained during the test.

The measured radiation patterns shown in Figures 15.8 and 15.9 confirm a
realized gain of ~16.2 dBi at 3 GHz, which is in excellent agreement with the

(a)

(b)   (c)

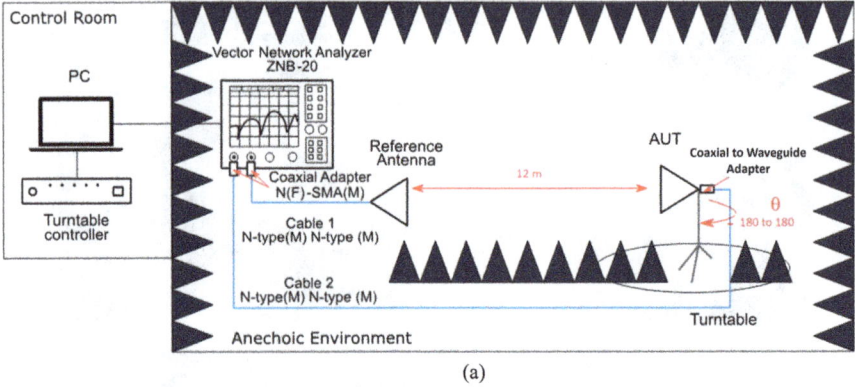

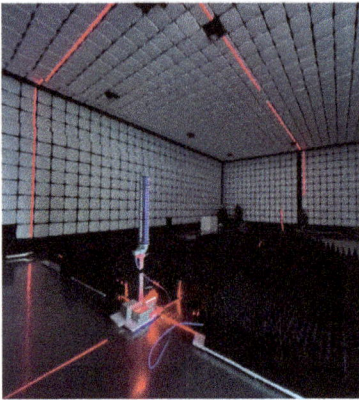

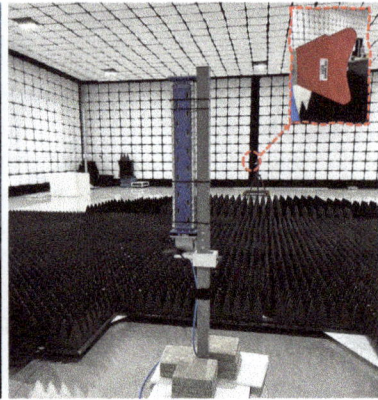

Figure 15.7   Antenna measurement setup in the semi-anechoic chamber.
(a) Schematic block diagram. (b) Laser alignment process. (c) AUT
and reference antenna placement.

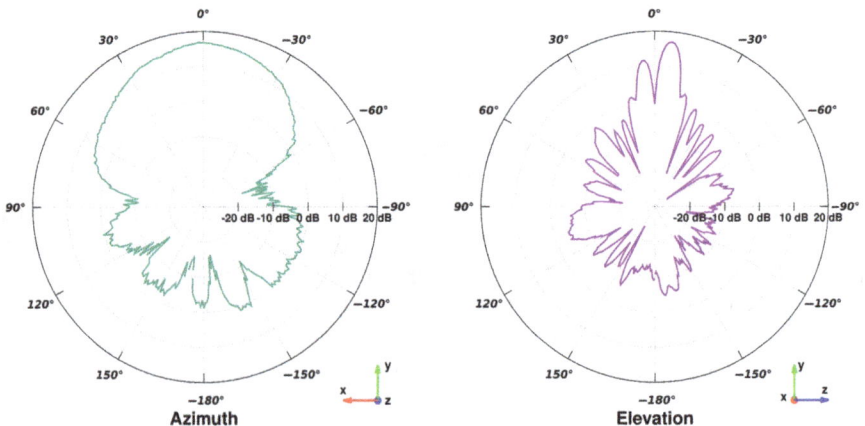

Figure 15.8   Measured 2D-radiation patterns of the proposed HPSWA

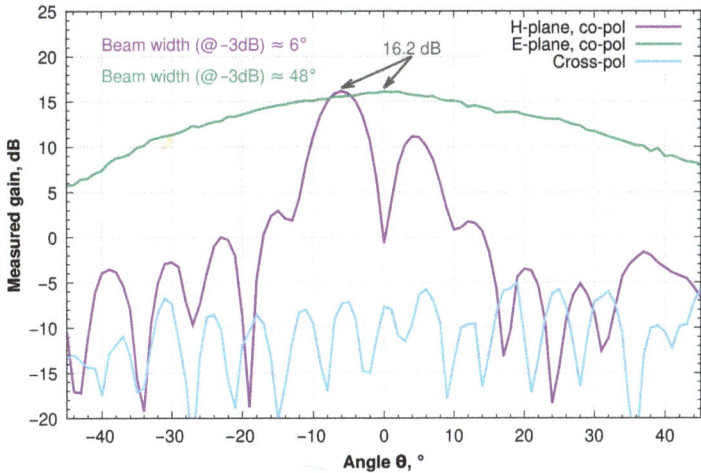

Figure 15.9   Co- and cross-polarization performance of the proposed HPSWA

16.6 dBi obtained in the simulation analysis (see Figure 15.3 for a reference). The measured -3 dB beamwidths are 48° and 6°, for azimuth and elevation planes respectively. It is worth mentioning that the antenna has been intentionally designed to radiate at a slanted angle of $\sim\!-6°$ in elevation (H-plane) as a predefined requirement.

15.2.5   High power testing

A narrow-band HPM generator, working in S-band, capable of generating multi-micro micro-second microwave pulses of up to 4 MW peak power was used for the high-power testing of the prototyped HPSWA. A scheme of the experiment setup is shown in Figure 15.10. The source, the HPSWA antenna, and the measurement sensors were placed in the semi-anechoic chamber, as shown in Figure 15.11. A ridged horn antenna and a free-field electric field D-dot sensor were used in the receiving part, 8.9 m away from the HPSWA. All the signals were recorded by a high-speed, 6 GHz, oscilloscope as shown in the inset of Figure 15.11.

During the experiments, the source was typically operated at a pulse repetition rate of 500 Hz, and the duration of each pulse was about 2 µs; the peak generated power was varied from 2.7 MW to 3.85 MW. Before switching the source on, the HPSWA was filled and set to 2 bars of SF6 to avoid breakdown inside and in the vicinity of the slots. Figure 15.12 shows a comparison of the measured electric field strength and the radiation pattern, measured and described in Section 15.2.4. The maximum of radiation using the high-power source matches the peak lobe angle in elevation, obtained during the radiation pattern characterization (i.e., $\sim\!-6°$ in elevation). The generator was set at 2.7 MW during this experiment.

Further high-power experiments were performed with the sensors oriented in the direction of the maximum radiation from the HPSWA. Figure 15.13 shows the measured electric field strength level for different transmitting power values, recorded in

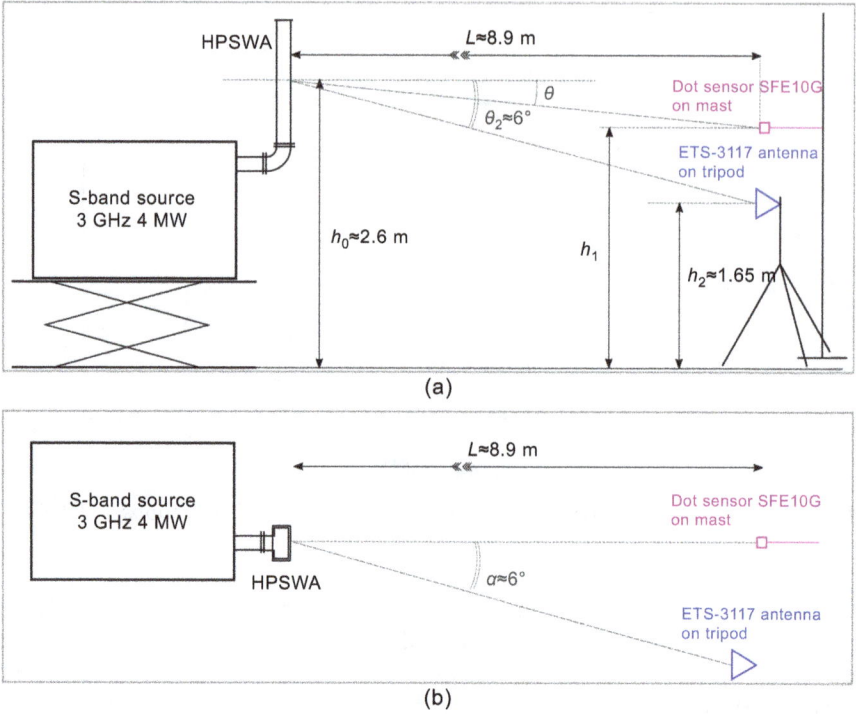

Figure 15.10   High-power test configuration: (a) side view and (b) top view

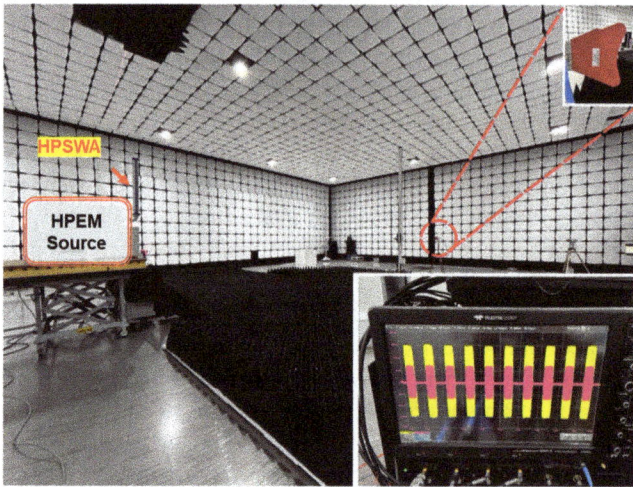

Figure 15.11   High-power test setup. In the bottom right corner, the typical waveforms registered by the oscilloscope during repetition-rate source operation are shown.

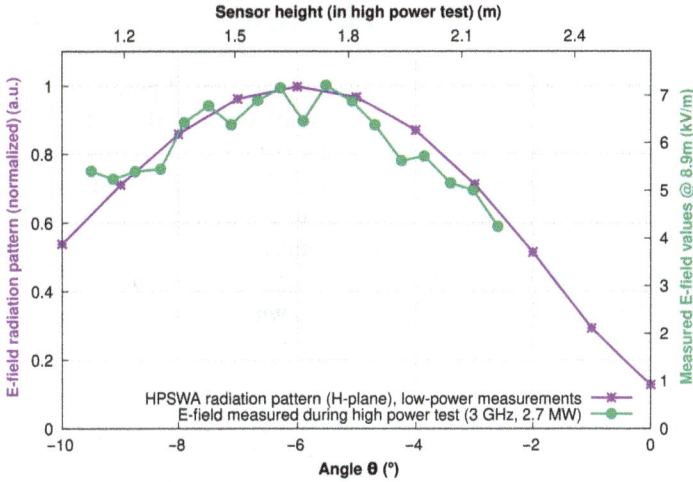

Figure 15.12   Measured radiation pattern of the HPSWA vs. the electric field strength peak amplitudes, from the high-power tests

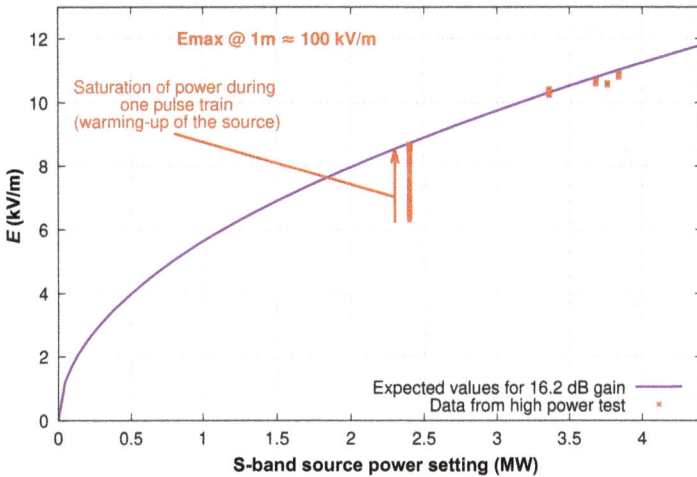

Figure 15.13   Electric field peak amplitude measured at 8.9 m from the HPSWA

pulse burst mode. The dynamic response of the generator is observed if the peak electric field is plotted, especially at low power levels, as seen in Figure 15.13, for 2.7 MW output power. For all the output power levels in the experiment, the measured electric field agrees well with the theoretical predictions for an antenna with 16.2 dB gain. The maximum electric field amplitude registered with the generator at maximum power (i.e., 4 MW) was about 100 kV/m, reduced to 1 m distance, i.e., the far-voltage figure of merit of the HPM radiator system: generator+HPSWA.

15.3   2D slotted waveguide array antenna

In an attempt to improve the equivalent isotropic radiated power (EIRP) figure of merit of the system, and part of a future line of work in the field of high-power slotted waveguide arrays, a 2D version of the HPSWA has been designed and the preliminary results presented in this section. This 2D-HPSWA has been designed for operation at 3 GHz as well. It employs a standard WR 284 waveguide as a base waveguide structure, feeding 6 by 12 horizontally oriented radiating slot elements. A CAD model of the proposed 2D-HPSWA is shown in Figure 15.14.

The simulated reflection coefficient is shown in Figure 15.15. An impedance bandwidth of 63 MHz was obtained. The radiation pattern is shown in Figure 15.16.

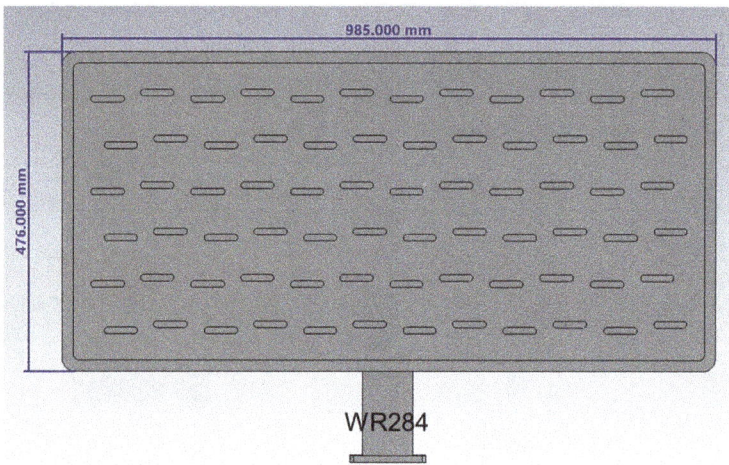

Figure 15.14   2D-HPSWA model for full-wave electromagnetic simulation

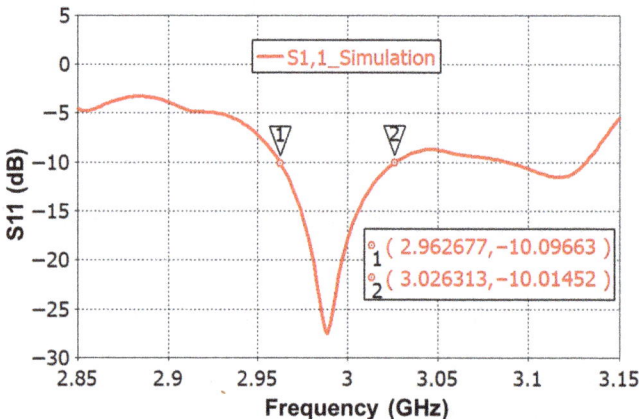

Figure 15.15   Simulated reflection coefficient of the 2D-HPSWA

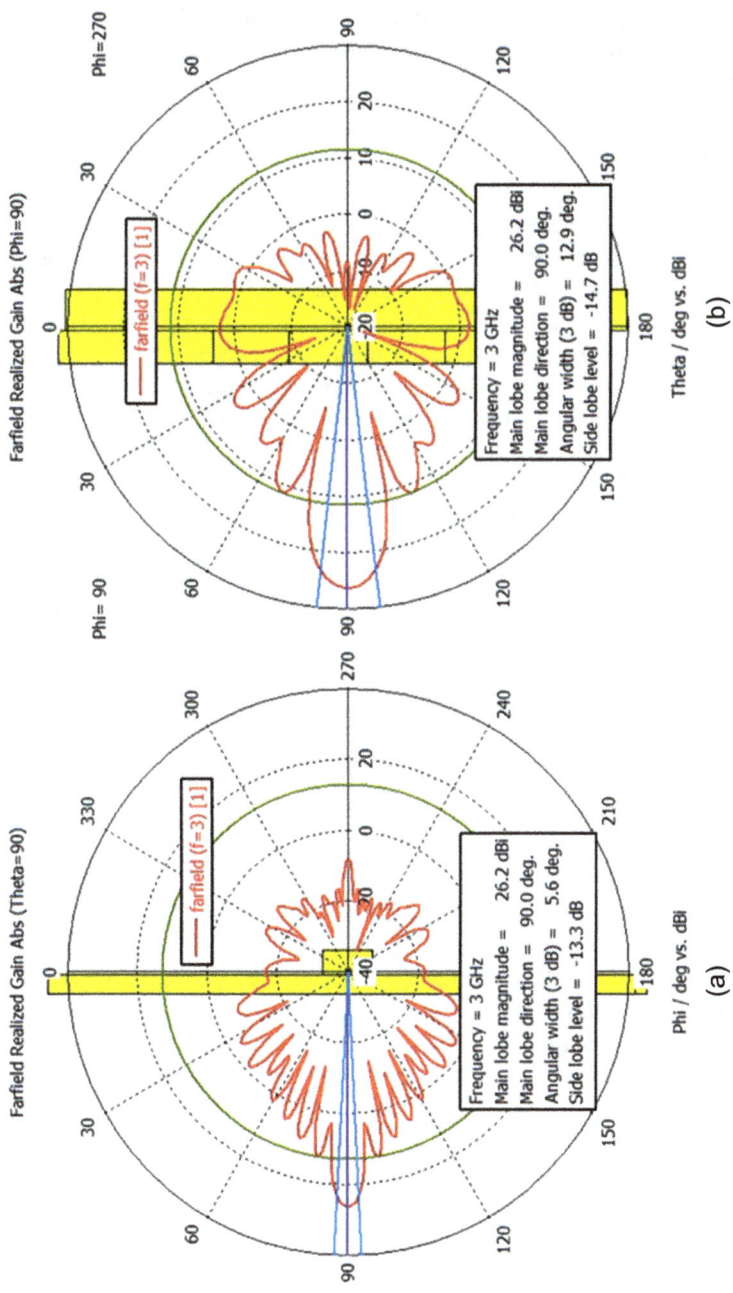

Figure 15.16 Simulated radiation pattern: (a) azimuth plane, and (b) elevation plane

An antenna gain of 26 dBi is observed on boresight. Please note that this 2D-HPSWA has been designed to have the maximum radiation intensity on boresight. Work is in progress to manufacture the antenna and conduct the structural and pneumatic testing, prior to the integration and subsequent high-power test with the S-band generator.

15.4   Conclusion

In this chapter, the design, manufacture, and characterization of the High-Power Slotted Waveguide Antenna (HPSWA) has been presented. The presented development marks a significant leap forward in antenna technology, specifically tailored for implementing multi-MW HPM Radiators in the S-band. By integrating advanced design principles and utilizing sulfur hexafluoride (SF6) as a pressurizing medium, the HPSWA effectively addresses the challenges of high-power transmission while ensuring operational integrity under pressurized media conditions. The extensive performance evaluations demonstrate a significantly high antenna gain and radiation characteristics in a compact form factor. The findings underscore the HPSWA's potential to meet the increasing demands of modern communications, radar, IEMI, and directed energy systems. The 2D-HPSWA has been presented as an alternative to increasing the equivalent radiated power of the HPM radiator by a factor of 10 while maintaining a compact planar design. As research progresses, future enhancements in designing and exploration of diverse operational frequencies will further broaden the scope of the HPSWA's applicability, positioning it as a key component in the evolution of antenna technologies for high-power radiation applications.

References

[1]   Josefsson L, and Rengarajan SR. *Slotted Waveguide Array Antennas: Theory, Analysis and Design*. SciTech Publishing; 2018.

[2]   Rahmat-Samii Y, Duan DW, Giri D, *et al.* Canonical examples of reflector antennas for high-power microwave applications. *IEEE Transactions on Electromagnetic Compatibility*. 1992;34(3):197–205.

[3]   Liang Y, Zhang J, Liu Q, *et al.* High-power dual-branch helical antenna. *IEEE Antennas and Wireless Propagation Letters*. 2018;17(3):472–475.

[4]   Jouade A, Himdi M, Chauloux A, *et al.* Mechanically pattern-reconfigurable bended horn antenna for high-power applications. *IEEE Antennas and Wireless Propagation Letters*. 2016;16:457–460.

[5]   Banelli A, AlEissaee N, Almansoori M, *et al.* A multi-megawatt range, dual-band waveguide antenna system. In: 2023 *International Microwave and Antenna Symposium (IMAS)*. Piscataway, NJ: IEEE; 2023. pp. 247–250.

[6]   Banelli A, Prado L, Vega F, *et al.* A dual-band radiating system for high-power microwave applications. In: *2023 IEEE International Symposium on*

Antennas and Propagation and USNC-URSI Radio Science Meeting (USNC-URSI). Piscataway, NJ: IEEE; 2023. pp. 1305–1306.

[7]   El Misilmani HM, Al-Husseini M, and Mervat M. Design of slotted waveguide antennas with low sidelobes for high power microwave applications. *Progress in Electromagnetics Research C*. 2015;56:15–28.

[8]   Tyagi Y, Mevada P, Chakrabarty S, *et al*. Broadband slotted waveguide array antenna with novel impedance matching network. In: *2019 IEEE Indian Conference on Antennas and Propagation (InCAP)*. Piscataway, NJ: IEEE; 2019. pp. 1–5.

[9]   Alhammadi H, Alyousef M, Alebri A, *et al*. Experimental investigation of the mixture of sulfur hexafluoride and nitrogen in pulsed regime. In: *2023 XXXVth General Assembly and Scientific Symposium of the International Union of Radio Science (URSI GASS)*. Piscataway, NJ: IEEE; 2023. pp. 1–2.

[10] Marcuvitz N. *Waveguide Handbook*. 21. Stevenage: IET; 1951.

[11] El Misilmani HM, Al-Husseini M, and Kabalan KY. Design procedure for planar slotted waveguide antenna arrays with controllable sidelobe level ratio for high power microwave applications. *Engineering Reports*. 2020;2(10): e12255.

Index

www.ingramcontent.com/pod-product-compliance
Lightning Source LLC
Chambersburg PA
CBHW050515190326
41458CB00005B/1548